新世纪计算机基础教育丛书　　丛书主编　谭浩强

Java程序设计
题解与上机指导（第四版）

辛运帏　饶一梅　编著

清华大学出版社
北京

内容简介

本书是和《Java 程序设计(第四版)》(辛运帏、饶一梅编著,清华大学出版社出版)配套使用的参考书,由题解和上机指导两部分组成。本书对《Java 程序设计(第四版)》中 13 章的全部习题给出了完整的解答。书中对论述题给出了简单的答案;对编程题给出了简单分析,论述了设计思路,并给出了相应的程序代码。这些代码均在 Java 8.0 环境下调试通过,程序运行结果以截图的方式提供给读者,供读者参考。在有些习题的后面,又给出了若干拓展思考题,可帮助读者进一步理解相关的习题。此外,本书精选了 9 个实验题目,供教师配合授课使用。

通过学习《Java 程序设计(第四版)》并配合本书的使用,能使读者更深入地了解 Java 语言,熟练掌握它,并能使用该语言编程完成特定的任务。

本书概念清晰、实用性强,可供学习 Java 语言的读者参考使用。

本书封面贴有清华大学出版社防伪标签,无标签者不得销售。
版权所有,侵权必究。举报: 010-62782989, beiqinquan@tup.tsinghua.edu.cn。

图书在版编目(CIP)数据

Java 程序设计题解与上机指导/辛运帏,饶一梅编著. —4 版. —北京: 清华大学出版社,2017
(2022.1重印)
(新世纪计算机基础教育丛书/谭浩强主编)
ISBN 978-7-302-47826-3

Ⅰ. ①J… Ⅱ. ①辛… ②饶… Ⅲ. ①JAVA 语言—程序设计 Ⅳ. ①TP312.8

中国版本图书馆 CIP 数据核字(2017)第 170461 号

责任编辑:焦　虹
封面设计:傅瑞学
责任校对:徐俊伟
责任印制:沈　露

出版发行:清华大学出版社
　　网　　　址:http://www.tup.com.cn, http://www.wqbook.com
　　地　　　址:北京清华大学学研大厦 A 座　　邮　编:100084
　　社　总　机:010-62770175　　　　　　　　　邮　购:010-83470235
　　投稿与读者服务:010-62776969, c-service@tup.tsinghua.edu.cn
　　质　量　反　馈:010-62772015, zhiliang@tup.tsinghua.edu.cn
　　课　件　下　载:http://www.tup.com.cn, 010-83470236
印 装 者:三河市铭诚印务有限公司
经　　销:全国新华书店
开　　本:185mm×260mm　　印　张:22.25　　字　数:531 千字
版　　次:2003 年 7 月第 1 版　　2017 年 9 月第 4 版　　印　次:2022 年 1 月第 6 次印刷
定　　价:59.00 元

产品编号:073892-02

丛书序言

现代科学技术的飞速发展,改变了世界,也改变了人类的生活。作为新世纪的大学生,应当站在时代发展的前列,掌握现代科学技术知识,调整自己的知识结构和能力结构,以适应社会发展的要求。新世纪需要具有丰富的现代科学知识,能够独立完成面临的任务,充满活力,有创新意识的新型人才。

掌握计算机知识和应用,无疑是培养新型人才的一个重要环节。现在计算机技术已深入到人类生活的各个角落,与其他学科紧密结合,成为推动各学科飞速发展的有力的催化剂。无论学什么专业的学生,都必须具备计算机的基础知识和应用能力。计算机既是现代科学技术的结晶,又是大众化的工具。学习计算机知识,不仅能够掌握有关知识,而且能培养人们的信息素养。这是高等学校全面素质教育中极为重要的一部分。

高校计算机基础教育应当遵循的理念是:面向应用需要;采用多种模式;启发自主学习;重视实践训练;加强创新意识;树立团队精神,培养信息素养。

计算机应用人才队伍由两部分人组成:一部分是计算机专业出身的计算机专业人才,他们是计算机应用人才队伍中的骨干力量;另一部分是各行各业中应用计算机的人员。这后一部分人一般并非计算机专业毕业,他们人数众多,既熟悉自己所从事的专业,又掌握计算机的应用知识,善于用计算机作为工具解决本领域中的问题。他们是计算机应用人才队伍中的基本力量。事实上,大部分应用软件都是由非计算机专业出身的计算机应用人员研制的,他们具有的这个优势是其他人难以代替的。从这个事实可以看到在非计算机专业中深入进行计算机教育的必要性。

非计算机专业中的计算机教育,无论目的、内容、教学体系、教材、教学方法等各方面与计算机专业有很大的不同,绝不能照搬计算机专业的模式和做法。全国高等院校计算机基础教育研究会自1984年成立以来,始终不渝地探索高校计算机基础教育的特点和规律。2004年,全国高等院校计算机基础教育研究会与清华大学出版社共同推出了《中国高等院校计算机基础教育课程体系2004》(简称CFC2004)。2006年、2008年又分别推出了《中国高等院校计算机基础教育课程体系2006》(简称CFC2006)及《中国高等院校计算机基础教育课程体系2008》(简称CFC2008),由清华大学出版社正式出版发行。

1988年起,我们根据教学实际的需要,组织编写了"计算机基础教育丛书",邀请有丰富教学经验的专家、学者先后编写了多种教材,由清华大学出版社出版。丛书出版后,迅速受到广大高校师生的欢迎,对高等学校的计算机基础教育起了积极的推动作用。广大读者反映这套教材定位准确,内容丰富,通俗易懂,符合大学生的特点。

1999年,根据新世纪的需要,在原有基础上组织出版了"新世纪计算机基础教育丛书"。由于内容符合需要,质量较高,被许多高校选为教材。丛书总发行量1000多万册,这在国内是罕见的。近期,我们又对丛书作了进一步的修订,根据发展的需要,增加了新的书目和内容。本丛书有以下特点:

(1) 内容新颖。根据21世纪的需要,重新确定丛书的内容,以符合计算机科学技术的发展和教学改革的要求。本丛书除保留了原丛书中经过实践考验且深受群众欢迎的优秀教材外,还编写了许多新的教材。在这些教材中反映了近年来迅速得到推广应用的一些计算机新技术,以后还将根据发展不断补充新的内容。

(2) 适合不同学校组织教学的需要。本丛书采用模块形式,提供了各种课程的教材,内容覆盖了高校计算机基础教育的各个方面。丛书中既有理工类专业的教材,也有文科和经济类专业的教材;既有必修课的教材,也包括一些选修课的教材。各类学校都可以从中选择到合适的教材。

(3) 符合初学者的特点。本丛书针对初学者的特点,以应用为目的,以应用为出发点,强调实用性。本丛书的作者都是长期在第一线从事高校计算机基础教育的教师,对学生的基础、特点和认识规律有深入的研究,在教学实践中积累了丰富的经验。可以说,每一本教材都是他们长期教学经验的总结。在教材的写法上,既注意概念的严谨和清晰,又特别注意采用读者容易理解的方法阐明看似深奥难懂的问题,做到例题丰富,通俗易懂,便于自学。这一点是本丛书一个十分重要的特点。

(4) 采用多样化的形式。除了教材这一基本形式外,有些教材还配有习题解答和上机指导,并提供电子教案。

总之,本丛书的指导思想是内容新颖、概念清晰、实用性强、通俗易懂、教材配套,简单概括为"新颖、清晰、实用、通俗、配套"。我们经过多年实践形成的这一套行之有效的创作风格,相信会受到广大读者的欢迎。

本丛书多年来得到了各方面人士的指导、支持和帮助,尤其是得到了全国高等院校计算机基础教育研究会的各位专家和各高校老师们的支持和帮助,我们在此表示由衷的感谢。本丛书肯定有不足之处,希望得到广大读者的批评指正。

丛 书 主 编
全国高等院校计算机基础教育研究会荣誉会长
谭 浩 强

前言

Java语言自问世以来,经历了多次版本升级,从安全机制、语法成分到API函数都有较大的修改。《Java程序设计》自2001年9月由清华大学出版社出版至今,经过了多次修订,目前已经出版了第4版。为配合使用《Java程序设计(第四版)》,我们编写了本书,旨在帮助读者检查Java语言的学习效果,尽快掌握Java语言。

本书对《Java程序设计(第四版)》中13章的全部习题给出了完整的解答。对论述题给出了简单的答案,内容主要摘自《Java程序设计(第四版)》。对编程题给出了对题目的简单分析,论述了设计思路,并给出了相应的程序代码。这些代码均在Java 8.0环境下调试通过,程序运行结果以截图的方式提供给读者,供读者参考。在有些习题的后面,又给出了若干拓展思考题,可帮助读者进一步理解相关的习题。

在《Java程序设计(第四版)》中,增加了一章新内容,即第7章Java语言的高级特性。相应地,本书也增加了一章,对教材中第7章的习题给出了简单解答。在高级特性中介绍了泛型及迭代器,所以对第三版中的程序代码也增加了对泛型及迭代器的使用,对相关程序做了修改及调试。

此外,本书精选了9个实验题目,可供教师配合授课使用。每个实验均列出了实验的目的,以此作为检查的目标。实验后可要求学生完成实验报告,让学生更加深入地理解Java语言,确保学生掌握编程技巧。

众所周知,大部分习题的解答不具有唯一性;特别是程序设计题目,给读者发挥潜能的余地非常大。本书中给出的这些解答和参考答案仅供参考,希望能起到抛砖引玉的作用。因为编者水平的局限性,书中的答案难免存在这样、那样的问题,实现的代码也不一定是最优的。读者可以参考本书中的内容和其他参考书中的内容,得出自己更全面的答案。至于程序代码,其实现的方式就更加多种多样,相信读者能在本书代码的基础之上,编写出功能更全面、效率更高的程序。

计算机技术是不断发展、不断完善的技术,Java语言也是如此。从诞生之日起,它的版本一直在更新中。就在本书编写过程及读者使用本书期间,相信Java又会有新的发展。读者应及时把握这些新动向,了解最新版本的相关信息,特别是及时更新自己机器上的JDK,以保持自己设计的代码与新版本同步。

　　本书是教学参考书，希望读者在使用、调试本书中代码的同时，既能加深对Java语言的理解，又能提高程序设计的能力，并在此过程中不断发现问题、思考问题、解决问题，把本书作为掌握知识的一个工具和桥梁。

　　本书由辛运帏、饶一梅编写，并运行通过了所有程序代码。

　　由于作者水平有限，对Java语言的掌握不够全面，书中难免有不妥之处，恳请广大读者特别是同行专家批评指正，在此我们表示深深的谢意。

编　者

于南开园

第3版前言

Java语言自问世以来,经历了多次版本升级,新版本从安全机制、语法成分到API函数都有较大的修改。《Java程序设计》自2001年9月由清华大学出版社出版至今,也经过了多次修订,目前已经出版了第三版。为配合使用《Java程序设计(第三版)》,我们编写了这本《Java程序设计题解与上机指导(第三版)》,旨在帮助读者检查Java语言的学习效果,尽快掌握Java语言。

本书对《Java程序设计(第三版)》一书中12章的全部习题做了完整解答。对论述题给出了简单的答案,内容主要摘自《Java程序设计(第三版)》。对编程题目给出了题目的简单分析,论述了设计思路,并给出了相应的程序代码。这些代码均在Java 5.0环境下调试通过,程序运行结果以截图的方式提供给读者,供读者参考。在有些习题的后面,又给出了若干拓展思考题,可帮助读者进一步理解相关的知识。此外,本书精选了9个实验题目,可供教师配合授课使用。每个实验均列出了实验的目的,以此作为检查的依据。实验后可要求学生完成实验报告,让学生更加深入理解Java语言,确保学生掌握编程技巧。

众所周知,大部分的习题解答不具有唯一性;特别是程序设计题目,给读者发挥潜能的余地非常大。本书中给出的这些解答和参考答案仅供参考,希望能起到抛砖引玉的作用。因为编者水平的局限性,书中的答案难免存在某些问题,实现的代码也不一定是最优的。读者可以参考本书中的内容和其他参考书中的内容,得出自己更全面的答案。至于程序代码,其实现的方式就更加多种多样,相信读者能在本书代码的基础之上,编写出功能更全面、效率更高的程序。

计算机技术是不断发展、不断完善的技术,Java语言也是如此。从诞生之日起,它的版本一直在更新中。就在本书编写过程及读者使用本书期间,相信Java语言又有了新的发展。读者应及时把握这些新动向,了解最新版本的相关信息,特别是及时更新自己机器上的JDK,以保持自己设计的代码与新版本同步。

本书是一本教学参考书,希望读者在使用、调试本书中代码的同时,既能加深对Java语言的理解,又能提高程序设计的能力,并在此过程中不断发现问题、思考问题、解决问题,把本书作为掌握知识的一个工具和桥梁。

本书由辛运帏、饶一梅编写,并运行通过了书中所有程序代码。

由于作者水平有限,对 Java 语言的掌握不够全面,书中难免有错误和不妥之处,恳请广大读者特别是同行专家批评指正,在此我们表示深深的谢意。

编　者
于南开园

目 录

第1部分 题 解

1 概述 —————————————————————————— 3

2 标识符、关键字和数据类型 ———————————— 19

3 表达式和流程控制语句 ——————————————— 46

4 数组和字符串 ———————————————————— 73

5 对象和类的进一步介绍 ——————————————— 93

6 Java 语言中的异常 ————————————————— 143

7 Java 语言的高级特性 ———————————————— 150

8	Java 的图形用户界面设计	157
9	Swing 组件	174
10	Java Applet	190
11	Java 数据流	205
12	线程	243
13	Java 的网络功能	252

第 2 部分　上 机 指 导

14	实验 1　熟悉系统及环境	261
15	实验 2　简单的输入/输出处理	263
16	实验 3　类的练习	267

17	实验 4　模拟彩票开奖游戏	273
18	实验 5　模拟 CD 出租销售店	281
19	实验 6　计算器	300
20	实验 7　用户界面设计	313
21	实验 8　多线程练习	332
22	实验 9　文件读写练习	338

第1部分

题　　解

第1章 概　　述

1.1 简述 Java 语言的特点。

解：Java 是简单的、面向对象的语言,具有分布性、安全性和健壮性等特点,1995 年由美国 Sun 公司向公众推出。具体地说,Java 语言有如下显著的特点。

1. 语法简单,功能强大

Java 语言的语法非常像 C++,同时去掉了 C++ 中不常用且容易出错的地方。例如,Java 中没有指针、结构等概念,没有预处理器,程序员不必自己释放占用的内存空间,因此在一定程度上减少了因内存混乱而导致的系统崩溃。另外,Java 强调其面向对象的特性,用它可以编制出非常复杂的系统。Java 不仅具备强大的功能,而且其解释器只占用很少的内存,适合在各种类型的机器上运行。

2. 分布式与安全性

Java 强调网络特性,内置了 TCP/IP、HTTP、FTP 协议类库,具备强大且易于使用的网络能力,便于开发网上应用系统。

Java 程序的三级代码安全检查机制可以有效地防止非法代码的侵入,阻止对内存的越权访问,能够避免病毒的侵害,成为 Internet 上最安全的技术之一。其新的安全机制在具备足够的安全特性的基础之上,又给了程序设计人员一定的自由度,允许他们设计功能更加强大的程序。

3. 与平台无关

提到 Java,有一句著名的口号:一次编写,到处运行。Java 规定了统一的数据类型,它们在任何机器上占用的内存大小都是不变的。Java 编译器将 Java 程序编译成二进制代码,即字节码(bytecode),这是 Java 源代码程序的底层表示。运行时环境针对不同的处理器指令系统,把字节码转换为不同的具体指令,因此 Java 可以跨平台使用,特别适合于网络应用,同时也为 Java 程序跨平台的无缝移植提供了很大的便利。

4. 解释编译两种运行方式

Java 程序可以经解释器得到字节码,所生成的字节码经过了精心设计,并进行了优化,因此运行速度较快,突破了以往解释性语言运行效率低的瓶颈。在现在的 Java 版本中又加入了编译功能(即 just-in-time 编译器,简称 JIT 编译器),生成器将字节代码转换成本机的机器代码,然后可以以较高速度执行,使得执行效率大幅度提高,达到了编译语言的水平。

5. 多线程

Java 内置了语言级多线程功能,可使用户程序并行执行。Java 提供的同步机制可保证各线程对共享数据的正确操作,完成各自的特定任务。在硬件条件允许的情况下,这些线程可以直接对应到各个 CPU 上,充分发挥硬件性能,减少用户等待时间。

6. 动态执行

Java 执行代码是在运行时动态载入的,程序可以自动进行版本升级,在网络环境下,可用于瘦客户机架构,减少维护工作。另外,类库中增加的新方法和其他实例,不会影响到原有程序的执行。

7. 丰富的 API 文档和类库

Java 为用户提供了详尽的 API 文档说明。Java 开发工具包中的类库包罗万象,应有尽有,程序员的开发工作可以在一个较高的层次上展开。这也正是 Java 受欢迎的重要原因之一。

1.2 什么是 Java 虚拟机?它包括哪几部分?

解:Java 虚拟机(Java virtual machine,JVM)规范中给出了它的定义:JVM 是在一台真正的机器上用软件方式实现的一台假想机。Java 虚拟机是运行 Java 程序必不可少的机制,它是编译后的 Java 程序和硬件系统之间的接口,程序员可以把 JVM 看作一个虚拟的处理器。编译后的 Java 程序指令并不直接在硬件系统的 CPU 上执行,而是由 JVM 执行。它不仅解释执行编译后的 Java 指令,而且还进行安全检查。它是 Java 程序能在多平台间进行无缝移植的可靠保证,同时也是 Java 程序的安全检验引擎。

JVM 的具体实现包括指令集(等价于 CPU 的指令集)、寄存器组、类文件格式、栈、垃圾收集堆、内存区。

1.3 简述 JVM 的工作机制。

解:JVM(Java 虚拟机)是运行 Java 程序必不可少的机制。编译后的 Java 程序指令并不直接在 CPU 上执行,而是由 JVM 执行。JVM 是编译后的 Java 程序和硬件系统之间的接口,程序员可以把 JVM 看作一个虚拟的处理器。它不仅解释执行编译后的 Java 指令,而且还进行安全检查。它是 Java 程序能在多平台间进行无缝移植的可靠保证,同时也是 Java 程序的安全检验引擎。

JVM 是在一台真正的机器上用软件方式实现的一台假想机。JVM 使用的代码存储在 .class 文件中。JVM 的部分指令很像真正的 CPU 指令,包括算术运算、流控制和数组元素访问等。

Java 虚拟机规范提供了编译所有 Java 代码的硬件平台。因为编译是针对虚拟机的,所以该规范能让 Java 程序独立于平台。它适用于每个具体的硬件平台,以保证为 JVM 编译的代码的运行。JVM 不但可以用软件实现,而且可以用硬件实现。

JVM 的代码格式为压缩的字节码,效率较高。由 JVM 字节码表示的程序必须保持原来的类型规定。Java 主要的类型检查是在编译时由字节码校验器完成的。Java 的任何解释器必须能执行符合 JVM 定义的类文件格式的任何类文件。

Java 虚拟机规范对运行时数据区域的划分及字节码的优化并不做严格的限制,它们的实现依平台的不同而有所不同。

1.4 Java 语言的安全机制有哪些?

解:在 Java 的不同版本中,都有不同的安全机制。在最初的 JDK 1.0 版本中,安全模型是所谓的"沙箱"模型,从网络上下载的代码只能在一个受限的环境中运行,这个环境像个箱子一样限制了代码能访问的资源。

在随后提出的 JDK 1.1 版本中,提出了"签名 Applet"的概念。有正确签名的 Applet 视同本地代码一样,可以使用本地的资源。没有签名的 Applet 还与前一版本一样,只在沙箱中运行。

在 Java 2 平台下,安全机制又有较大改善。它允许用户自己设定相关的安全级别。另外,对于应用程序,也采取了和 Applet 一样的安全策略,程序员可以根据需要对本地代码或是远程代码进行设定,以保证程序更加安全、高效地运行。

在 Java 程序环境中,重要的几个组成部分包括 Java 解释器、类下载器及字节码校验器。

1. Java 解释器

Java 解释器只能执行为 JVM 编译的代码。Java 解释器有三项主要工作:

(1) 下载代码——由类下载器完成。
(2) 校验代码——由字节码校验器完成。
(3) 运行代码——由运行时解释器完成。

2. 类下载器

Java 运行时系统区别对待来自不同源的类文件。它可能从本地文件系统中下载类文件,也可能从 Internet 上使用类下载器下载类文件。运行时系统动态决定程序运行时所需的类文件,并把它们下载到内存中,将类、接口与运行时系统相连接。类下载器把本地文件系统的类名空间和网络源输入的类名空间区分开来,以增加安全性。因为内置的类总是先被检查,所以可以防止起破坏作用的应用程序的侵袭。

所有的类下载完毕后,开始确定可执行文件的内存分配。此时,指定具体的内存地址,并创建查询表。因为内存分配是在运行时进行的,并且 Java 解释器可阻止访问可能给操作系统带来破坏的非法代码地址,所以增加了保护性。

3. 字节码校验器

Java 代码在机器上真正执行前要经过几次测试。程序通过字节码校验器检查代码的安全性,字节码校验器检测代码段的格式,并严格按照规则来检查非法代码段——伪造的指针、对目标的访问权限违例或是试图改变目标类型或类的代码。通过网络传送的所有类文件都要经过字节码校验器的检验。

字节码校验器要对程序中的代码进行 4 趟扫描。这可以保证代码依从 JVM 规范,并且不破坏系统的完整性。校验器主要检查以下几项内容:

(1) 类遵从 JVM 的类文件格式。
(2) 不出现访问违例情况。
(3) 代码不会引起运算栈溢出。
(4) 所有运算代码的参数类型总是正确的。
(5) 不会发生非法数据转换,如把整数转换为指针。
(6) 对象域访问是合法的。

如果完成所有的扫描之后不返回任何错误信息,就可以保证 Java 程序的安全性了。

1.5 Java 的垃圾收集机制与其他语言相比有什么特点?

解:许多程序设计语言允许在程序运行时动态分配内存,并将所分配的内存块的开始地址以指针的形式返回给程序,供程序员使用。一旦不再需要所分配的内存,程序或运行时环境最好将内存释放,避免内存越界时得到意外结果。

在 C 和 C++ 以及其他许多语言中,由程序开发人员负责内存的释放,但程序开发人员并不总是清楚内存应该在何时释放。如果不及时释放不再需要的内存,则可能会因内存资源的枯竭而使程序不能继续执行;另一方面,如果释放了仍在使用的内存,可能导致程序崩溃或系统混乱。这些不能正确使用内存的程序被称为有"内存漏洞"。

在 Java 中,程序员不必亲自释放内存,它提供了后台系统级线程,记录每次内存分配的情况,并统计每个内存指针的引用次数。引用次数为 0 则表示这块内存不再使用。在 Java 虚拟机运行时环境闲置时,垃圾收集线程将检查是否存在引用次数为 0 的内存指针;如果有,垃圾收集线程则把该内存"标记"为"清除"(释放)。

在 Java 程序生存期内,垃圾收集将自动进行,无需用户释放内存,从而消除了内存漏洞。

1.6 上机调试主教材第 1 章 1.2 节中的程序 1-1,直至得到正确结果。

解:程序中,第 6 行代码将在屏幕上显示一条信息:

Hello World!

使用编辑器(例如记事本)编辑程序 1-1,并保存之。文件名为 HelloWorldApp.java,保存文件时注意文件名的大小写。然后使用命令行 javac HelloWorldApp.java 编译程序,得到类文件 HelloWorldApp.class;使用命令行 java HelloWorldApp 执行该程序。执行结果如图 1-1 所示。

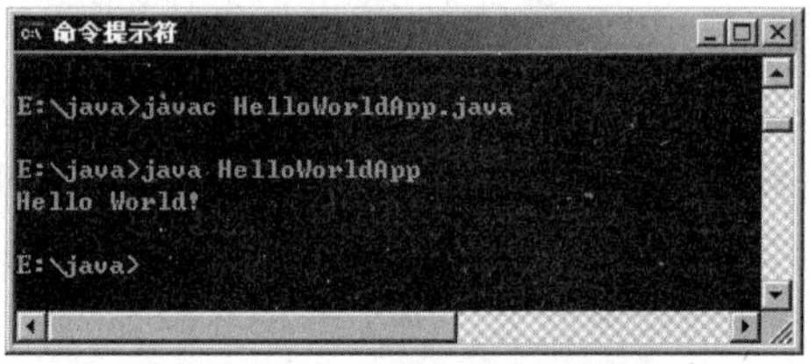

图 1-1 程序 1-1 的执行结果

【拓展思考】

(1) 调试成功后,按下列步骤对程序进行修改。每次修改一个,记录下编译程序产生的错误信息。每次加入新的错误之前修正前一个错误。如果没有产生错误信息,请解释原因。在每次修改前试着预测会出现什么情况。

① 将第 4 行的 HelloWorldApp 改为 HelloworldApp。

② 将第 6 行的 Hello World 改为 hello world。

③ 删除第 6 行字符串中第一个引号。

④ 删除第 6 行字符串中最后一个引号。

⑤ 将 main 改为 man。

⑥ 将 println 改为 bogus。

⑦ 删除 println 语句最后的分号。

⑧ 删除程序中最后一个括号。

(2) 下述情形哪些属于编译时错误?哪些属于运行时错误?哪些属于逻辑错误?

① 想对两个数进行加法运算时却进行了乘法运算。

② 表达式中被 0 除。

③ 程序语句的最后忘记写分号。

④ 在输出时拼错了一个字。

⑤ 得到一个不准确的结果。

⑥ 应该输入"("却输入了"{"。

1.7 练习使用浏览器查看 Java API 文档。

解：目前大多数读者使用的工作环境是 Windows 系列操作系统，例如 Windows 7 或 Windows 10 等，所以本书以 Windows 环境为平台，使用 JDK SE 8 版本(jdk-8u131-docs-all)。

假设读者目前的系统中已安装 JDK 8 文档，找到文件 index.html，双击来启动它。此时出现如图 1-2 所示的界面窗口。

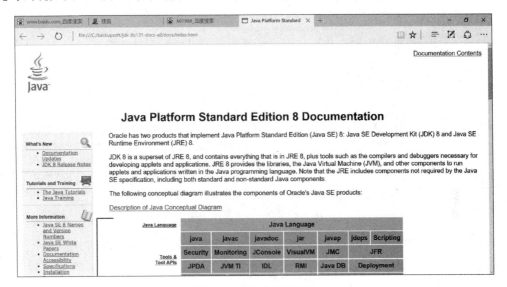

图 1-2　JDK 的初始界面

单击 Java SEAPI 进入类文档窗口，如图 1-3 所示。

图 1-3　API 文档

这个窗口分为三部分：左上部分显示Java JDK 8中提供的所有包的信息。选中某个包后，将在左下部分显示这个包中所有接口及类的信息。例如，选择查看java.lang包，窗口显示的内容如图1-4所示。

图1-4　java.lang包内信息

如果想进一步查看包中Integer类的信息，选中Integer，右侧窗口部分将显示java.lang中Integer类的所有接口及类的内容，向下拉动滚卷条，定位到所需的位置就可以了。

一般地，一个类中的信息包括以下几部分：
- 字段摘要（Field Summary）；
- 构造方法摘要（Constructor Summary）；
- 方法摘要（Method Summary）；
- 字段详细信息（Field Detail）；
- 构造方法详细信息（Constructor Detail）；
- 方法详细信息（Method Detail）。

这6部分内容是成对出现的。摘要部分简单介绍相关的内容，对应的详细信息部分深入列出具体内容。

字段摘要中列出类中成员变量的信息，包括名字、类型及含义。字段详细信息中将详细介绍这些成员变量。

构造方法摘要中列出类构造方法的信息，包括参数列表并解释所创建的实例。构造方法的详细信息显示在构造方法详细信息部分中。

在方法摘要中可以查找到要使用的方法名，在方法详细信息中将详细介绍该方法的使用方法，包括调用参数表及返回值的情况。例如，图1-5所示的窗口中显示的是toString()方法的详细情况，该方法将返回整数所对应的字符串。

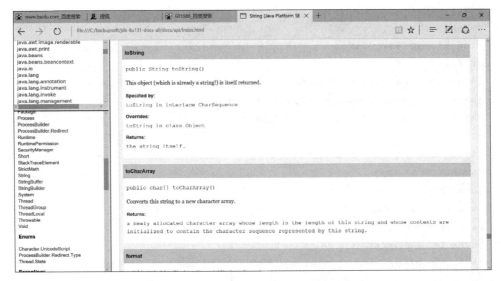

图 1-5 toString()方法的介绍

1.8 列出 Java API 文档中所有的包名。

解：进入 API 文档，可以查看预提供的所有的包的信息。Java API 文档中共包含 43 个大包，有些大包中又包含子包。所有包名如下：

java.applet；

java.awt、java.awt.color、java.awt.datatransfer、java.awt.dnd、java.awt.event、java.awt.font、java.awt.geom、java.awt.im、java.awt.im.spi、java.awt.image、java.awt.image.renderable、java.awt.print；

java.beans、java.beans.beancontext；

java.io；

java.lang、java.lang.annotation、java.lang.instrument、java.lang.management、java.lang.ref、java.lang.reflect；

java.math；

java.net；

java.nio、java.nio.channels、java.nio.channels.spi、java.nio.charset、java.nio.charset.spi；

java.rmi、java.rmi.activation、java.rmi.dgc、java.rmi.registry、java.rmi.server；

java.security、java.security.acl、java.security.cert、java.security.interfaces、java.security.spec；

java.sql；

java.text；

java.util、java.util.concurrent、java.util.concurrent.atomic、java.util.concurrent.locks、java.util.jar、java.util.logging、java.util.prefs、java.util.regex、java.util.zip；

javax.accessibility；

javax.activity；

javax.crypto、javax.crypto.interfaces、javax.crypto.spec;
javax.imageio、javax.imageio.event、javax.imageio.metadata、javax.imageio.plugins.bmp、javax.imageio.plugins.jpeg、javax.imageio.spi、javax.imageio.stream;
javax.management、javax.management.loading、javax.management.modelmbean、javax.management.monitor、javax.management.openmbean、javax.management.relation、javax.management.remote、javax.management.remote.rmi、javax.management.timer;
javax.naming、javax.naming.directory、javax.naming.event、javax.naming.ldap、javax.naming.spi;
javax.net、javax.net.ssl;
javax.print、javax.print.attribute、javax.print.attribute.standard、javax.print.event;
javax.rmi、javax.rmi.CORBA、javax.rmi.ssl;
javax.security.auth、javax.security.auth.callback、javax.security.auth.kerberos、javax.security.auth.login、javax.security.auth.spi、javax.security.auth.x500、javax.security.cert、javax.security.sasl;
javax.sound.midi、javax.sound.midi.spi、javax.sound.sampled、javax.sound.sampled.spi;
javax.sql、javax.sql.rowset、javax.sql.rowset.serial、javax.sql.rowset.spi;
javax.swing、javax.swing.border、javax.swing.colorchooser、javax.swing.event、javax.swing.filechooser、javax.swing.plaf、javax.swing.plaf.basic、javax.swing.plaf.metal、javax.swing.plaf.multi、javax.swing.plaf.synth、javax.swing.table、javax.swing.text、javax.swing.text.html、javax.swing.text.html.parser、javax.swing.text.rtf、javax.swing.tree、javax.swing.undo;
javax.transaction、javax.transaction.xa;
javax.xml、javax.xml.datatype、javax.xml.namespace、javax.xml.parsers、javax.xml.transform、javax.xml.transform.dom、javax.xml.transform.sax、javax.xml.transform.stream、javax.xml.validation、javax.xml.xpath;
org.ietf.jgss;
org.omg.CORBA、org.omg.CORBA_2_3、org.omg.CORBA_2_3.portable、org.omg.CORBA.DynAnyPackage、org.omg.CORBA.ORBPackage、org.omg.CORBA.portable、org.omg.CORBA.TypeCodePackage; org.omg.CosNaming、org.omg.CosNaming.NamingContextExtPackage、org.omg.CosNaming.NamingContextPackage; org.omg.Dynamic、org.omg.DynamicAny、org.omg.DynamicAny.DynAnyFactoryPackage、org.omg.DynamicAny.DynAnyPackage; org.omg.IOP、org.omg.IOP.CodecFactoryPackage、org.omg.IOP.CodecPackage; org.omg.Messaging; org.omg.PortableInterceptor、org.omg.PortableInterceptor.ORBInitInfoPackage、org.omg.PortableServer、org.omg.PortableServer.CurrentPackage、org.omg.PortableServer.POAManagerPackage、org.omg.PortableServer.POAPackage、org.omg.PortableServer.portable、org.omg.PortableServer.ServantLocatorPackage; org.omg.SendingContext; org.omg.stub.java.rmi;
org.w3c.dom、org.w3c.dom.bootstrap、org.w3c.dom.events、org.w3c.dom.ls;
org.xml.sax、org.xml.sax.ext、org.xml.sax.helpers。

1.9 列出 java.lang 中所有的类。

解：进入 API 文档中，选中 java.lang 包，则包中所含的所有类如下所示：

Boolean、Byte、Character、Character.Subset、Character.UnicodeBlock、Class、ClassLoader、Compiler、Double、Enum、Float、InheritableThreadLocal、Integer、Long、Math、Number、Object、Package、Process、ProcessBuilder、Runtime、RuntimePermission、SecurityManager、Short、StackTraceElement、StrictMath、String、StringBuffer、StringBuilder、System、Thread、ThreadGroup、ThreadLocal、Throwable、Void。

1.10 列出 java.applet 中 Applet 类的所有父类。

解：进入 API 文档中可以看到，java.applet 中 Applet 类的所有父类如下：

java.lang.Object、java.awt.Component、java.awt.Container、java.awt.Panel，如图 1-6 所示。

```
java.lang.Object
  └─java.awt.Component
      └─java.awt.Container
          └─java.awt.Panel
              └─java.applet.Applet
```

图 1-6　Applet 类的所有父类

1.11 列出 java.awt 中的所有接口。

解：java.awt 中的所有接口包括：

ActiveEvent、Adjustable、Composite、CompositeContext、ItemSelectable、KeyEventDispatcher、KeyEventPostProcessor、LayoutManager、LayoutManager2、MenuContainer、Paint、PaintContext、PrintGraphics、Shape、Stroke、Transparency。

1.12 列出 java.lang.Math 类中的常用方法，并总结 Math 类的功能。

解：java.lang.Math 类中的常用方法如表 1-1 所示。

表 1-1　java.lang.Math 类中的常用方法

方　　法	返回值类型	描　　述
abs(double a)	static double	返回 double 值的绝对值
abs(float a)	static float	返回 float 值的绝对值
abs(int a)	static int	返回 int 值的绝对值
abs(long a)	static long	返回 long 值的绝对值
acos(double a)	static double	返回角的反余弦，范围在 0.0 到 π 之间
asin(double a)	static double	返回角的反正弦，范围在 -π/2 到 π/2 之间
atan(double a)	static double	返回角的反正切，范围在 -π/2 到 π/2 之间
atan2(double y, double x)	static double	将矩形坐标 (x,y) 转换成极坐标 (r,theta)
cbrt(double a)	static double	返回 double 值的立方根
ceil(double a)	static double	返回最小的（最接近负无穷大）double 值，该值大于或等于参数，并且等于某个整数
cos(double a)	static double	返回角的三角余弦

续表

方　　法	返回值类型	描　　述
cosh(double x)	static double	返回 double 值的双曲线余弦
exp(double a)	static double	返回欧拉数 e 的 double 次幂的值
expm1(double x)	static double	返回 e^x-1
floor(double a)	static double	返回最大的(最接近正无穷大)double 值,该值小于或等于参数,并且等于某个整数
hypot(double x,double y)	static double	返回 $sqrt(x^2+y^2)$,没有中间溢出或下溢
IEEEremainder(double f1, double f2)	static double	按照 IEEE 754 标准的规定,对两个参数进行余数运算
log(double a)	static double	返回(底数是 e)double 值的自然对数
log10(double a)	static double	返回 double 值的底数为 10 的对数
log1p(double x)	static double	返回参数与 1 的和的自然对数
max(double a,double b)	static double	返回两个 double 值中较大的一个
max(float a,float b)	static float	返回两个 float 值中较大的一个
max(int a,int b)	static int	返回两个 int 值中较大的一个
max(long a,long b)	static long	返回两个 long 值中较大的一个
min(double a,double b)	static double	返回两个 double 值中较小的一个
min(float a,float b)	static float	返回两个 float 值中较小的一个
min(int a,int b)	static int	返回两个 int 值中较小的一个
min(long a,long b)	static long	返回两个 long 值中较小的一个
pow(double a,double b)	static double	返回第一个参数的第二个参数次幂的值
random()	static double	返回带正号的 double 值,大于或等于 0.0,小于 1.0
rint(double a)	static double	返回其值最接近参数并且是整数的 double 值
round(double a)	static long	返回最接近参数的 long
round(float a)	static int	返回最接近参数的 int
signum(double d)	static double	返回参数的符号函数;如果参数是零,则返回零;如果参数大于零,则返回 1.0;如果参数小于零,则返回 -1.0
signum(float f)	static float	返回参数的符号函数;如果参数是零,则返回零;如果参数大于零,则返回 1.0;如果参数小于零,则返回 -1.0
sin(double a)	static double	返回角的三角正弦
sinh(double x)	static double	返回 double 值的双曲线正弦

续表

方 法	返回值类型	描 述
sqrt(double a)	static double	返回正确舍入的 double 值的正平方根
tan(double a)	static double	返回角的三角正切
tanh(double x)	static double	返回 double 值的双曲线余弦
toDegrees(double angrad)	static double	将用弧度测量的角转换为近似相等的用度数测量的角
toRadians(double angdeg)	static double	将用度数测量的角转换为近似相等的用弧度测量的角
ulp(double d)	static double	返回参数的 ulp 大小
ulp(float f)	static float	返回参数的 ulp 大小

可以看出，Math 类的功能就是执行基本的数学运算。

1.13 查阅 API 文档，列出 java.lang.String 类的常用方法。

解：java.lang.String 类的常用方法如表 1-2 所示。

表 1-2 java.lang.String 类的常用方法

方 法	返回值类型	描 述
charAt(int index)	char	返回指定索引处的 char 值
codePointAt(int index)	int	返回指定索引处的字符(Unicode 代码点)
codePointBefore(int index)	int	返回指定索引之前的字符(Unicode 代码点)
codePointCount(int beginIndex, int endIndex)	int	返回此 String 的指定文本范围中的 Unicode 代码点数
compareTo(String anotherString)	int	按字典顺序比较两个字符串
compareToIgnoreCase(String str)	int	不考虑大小写，按字典顺序比较两个字符串
concat(String str)	String	将指定字符串连接到此字符串的结尾
contains(CharSequence s)	boolean	当且仅当此字符串包含 char 值的指定序列时，才返回 true
contentEquals(CharSequence cs)	boolean	当且仅当此 String 表示与指定序列相同的 char 值时，才返回 true
contentEquals(StringBuffer sb)	boolean	当且仅当此 String 表示与指定的 StringBuffer 相同的字符序列时，才返回 true
copyValueOf(char[] data)	static String	返回指定数组中表示该字符序列的字符串
copyValueOf(char[] data, int offset, int count)	static String	返回指定数组中表示该字符序列的字符串
endsWith(String suffix)	boolean	测试此字符串是否以指定的后缀结束

续表

方　　法	返回值类型	描　　述
equals(Object anObject)	boolean	比较此字符串与指定的对象
equalsIgnoreCase(String anotherString)	boolean	将此 String 与另一个 String 进行比较,不考虑大小写
format(Locale l, String format, Object. args)	static String	使用指定的语言环境、格式字符串和参数返回一个格式化字符串
format(String format, Object. args)	static String	使用指定的格式字符串和参数返回一个格式化字符串
getBytes()	byte[]	使用平台默认的字符集将此 String 解码为字节序列,并将结果存储到一个新的字节数组中
getBytes(String charsetName)	byte[]	使用指定的字符集将此 String 解码为字节序列,并将结果存储到一个新的字节数组中
getChars(int srcBegin, int srcEnd, char[] dst, int dstBegin)	void	将字符从此字符串复制到目标字符数组
hashCode()	int	返回此字符串的哈希码
indexOf(int ch)	int	返回指定字符在此字符串中第一次出现处的索引
indexOf(int ch, int fromIndex)	int	从指定的索引开始搜索,返回在此字符串中第一次出现指定字符处的索引
indexOf(String str)	int	返回第一次出现的指定子字符串在此字符串中的索引
indexOf(String str, int fromIndex)	int	从指定的索引处开始,返回第一次出现的指定子字符串在此字符串中的索引
intern()	String	返回字符串对象的规范化表示形式
lastIndexOf(int ch)	int	返回最后一次出现的指定字符在此字符串中的索引
lastIndexOf(int ch, int fromIndex)	int	从指定的索引处开始进行后向搜索,返回最后一次出现的指定字符在此字符串中的索引
lastIndexOf(String str)	int	返回在此字符串中最右边出现的指定子字符串的索引
lastIndexOf(String str, int fromIndex)	int	从指定的索引处开始向后搜索,返回在此字符串中最后一次出现的指定子字符串的索引
length()	int	返回此字符串的长度
matches(String regex)	boolean	通知此字符串是否匹配给定的正则表达式

续表

方法	返回值类型	描述
offsetByCodePoints(int index, int codePointOffset)	int	返回此 String 中从给定的 index 处偏移 codePointOffset 个代码点的索引
regionMatches(boolean ignoreCase, int toffset, String other, int ooffset, int len)	boolean	测试两个字符串区域是否相等
regionMatches(int toffset, String other, int ooffset, int len)	boolean	测试两个字符串区域是否相等
replace(char oldChar, char newChar)	String	返回一个新的字符串,它是通过用 newChar 替换此字符串中出现的所有 oldChar 而生成的
replace(CharSequence target, CharSequence replacement)	String	使用指定的字面值替换序列替换此字符串匹配字面值目标序列的每个子字符串
replaceAll(String regex, String replacement)	String	使用给定的 replacement 字符串替换此字符串匹配给定的正则表达式的每个子字符串
replaceFirst(String regex, String replacement)	String	使用给定的 replacement 字符串替换此字符串匹配给定的正则表达式的第一个子字符串
split(String regex)	String[]	根据给定的正则表达式的匹配来拆分此字符串
split(String regex, int limit)	String[]	根据匹配给定的正则表达式来拆分此字符串
startsWith(String prefix)	boolean	测试此字符串是否以指定的前缀开始
startsWith(String prefix, int toffset)	boolean	测试此字符串是否以指定的前缀开始,该前缀以指定的索引开始
subSequence(int beginIndex, int endIndex)	CharSequence	返回一个新的字符序列,它是此序列的一个子序列
substring(int beginIndex)	String	返回一个新的字符串,它是此字符串的一个子字符串
substring(int beginIndex, int endIndex)	String	返回一个新字符串,它是此字符串的一个子字符串
toCharArray()	char[]	将此字符串转换为一个新的字符数组
toLowerCase()	String	使用默认语言环境的规则将此 String 中的所有字符都转换为小写
toLowerCase(Locale locale)	String	使用给定 Locale 的规则将此 String 中的所有字符都转换为小写
toString()	String	返回此对象本身(它已经是一个字符串!)
toUpperCase()	String	使用默认语言环境的规则将此 String 中的所有字符都转换为大写

续表

方法	返回值类型	描述
toUpperCase(Locale locale)	String	使用给定的 Locale 规则将此 String 中的所有字符都转换为大写
trim()	String	返回字符串的副本,忽略前导空白和尾部空白
valueOf(boolean b)	static String	返回 boolean 参数的字符串表示形式
valueOf(char c)	static String	返回 char 参数的字符串表示形式
valueOf(char[] data)	static String	返回 char 数组参数的字符串表示形式
valueOf(char[] data, int offset, int count)	static String	返回 char 数组参数的特定子数组的字符串表示形式
valueOf(double d)	static String	返回 double 参数的字符串表示形式
valueOf(float f)	static String	返回 float 参数的字符串表示形式
valueOf(int i)	static String	返回 int 参数的字符串表示形式
valueOf(long l)	static String	返回 long 参数的字符串表示形式
valueOf(Object obj)	static String	返回 Object 参数的字符串表示形式

1.14 查阅 API 文档,列出 java.util.Random 类的常用方法。

解:java.util.Random 类的常用方法如表 1-3 所示。

表 1-3 java.util.Random 类的常用方法

方法	返回值类型	描述
next(int bits)	protected int	生成下一个伪随机数
nextBoolean()	boolean	返回下一个伪随机数,它是从此随机数生成器的序列中取出的、均匀分布的 boolean 值
nextBytes(byte[] bytes)	void	生成随机字节并将其置于用户提供的字节数组中
nextDouble()	double	返回下一个伪随机数,它是从此随机数生成器的序列中取出的、在 0.0 和 1.0 之间均匀分布的 double 值
nextFloat()	float	返回下一个伪随机数,它是从此随机数生成器的序列中取出的、在 0.0 和 1.0 之间均匀分布的 float 值
nextGaussian()	double	返回下一个伪随机数,它是从此随机数生成器的序列中取出的、呈高斯("正常地")分布的 double 值,其平均值是 0.0,标准偏差是 1.0
nextInt()	int	返回下一个伪随机数,它是此随机数生成器的序列中均匀分布的 int 值
nextInt(int n)	int	返回一个伪随机数,它是从此随机数生成器的序列中取出的、在 0(包括)和指定值(不包括)之间均匀分布的 int 值

续表

方法	返回值类型	描述
nextLong()	long	返回下一个伪随机数，它是从此随机数生成器的序列中取出的、均匀分布的 long 值
setSeed(long seed)	void	使用单个 long 种子设置此随机数生成器的种子

1.15 查阅 API 文档，列出 java.awt.Color 类的常用方法。

解：java.awt.Color 类的常用方法如表 1-4 所示。

表 1-4 java.awt.Color 类的常用方法

方法	返回值类型	描述
brighter()	Color	创建一个新 Color，它具有比此 Color 更亮的颜色
createContext(ColorModel cm, Rectangle r, Rectangle2D r2d, AffineTransform xform, RenderingHints hints)	PaintContext	创建并返回用来生成固定颜色模式的 PaintContext
darker()	Color	创建一个新 Color，它具有比此 Color 更暗的颜色
decode(String nm)	static Color	将 String 转换成整数，并返回指定的不透明 Color
equals(Object obj)	boolean	确定另一个对象是否与此 Color 相同
getAlpha()	int	返回位于 0~255 中的 alpha 分量
getBlue()	int	返回默认 sRGB 空间中位于 0~255 中的蓝色分量
getColor(String nm)	static Color	查找系统属性中的一种颜色
getColor(String nm, Color v)	static Color	查找系统属性中的一种颜色
getColor(String nm, int v)	static Color	查找系统属性中的一种颜色
getColorComponents(ColorSpace cspace, float[] compArray)	float[]	根据由 cspace 参数指定的 ColorSpace，返回一个 float 数组，该数组只包含 Color 的颜色分量
getColorComponents(float[] compArray)	float[]	根据 Color 的 ColorSpace，返回一个 float 数组，该数组只包含 Color 的颜色分量
getColorSpace()	ColorSpace	返回此 Color 的 ColorSpace
getComponents(ColorSpace cspace, float[] compArray)	float[]	根据由 cspace 参数指定的 ColorSpace，返回一个 float 数组，该数组只包含 Color 的 alpha 分量

续表

方　　法	返回值类型	描　　述
getComponents(float[] compArray)	float[]	根据 Color 的 ColorSpace,返回一个 float 数组,该数组包含 Color 的颜色分量和 alpha 分量
getGreen()	int	返回默认 sRGB 空间中位于 0~255 中的绿色分量
getHSBColor(float h,float s,float b)	static Color	根据所指定的数值,创建一个基于 HSB 颜色模型的 Color 对象
getRed()	int	返回默认 sRGB 空间中位于 0~255 中的红色分量
getRGB()	int	返回默认 sRGB ColorModel 中表示颜色的 RGB 值
getRGBColorComponents(float[] compArray)	float[]	根据默认的 sRGB color space,返回一个 float 数组,该数组只包含 Color 的颜色分量
getRGBComponents(float[] compArray)	float[]	根据默认的 sRGB color space,返回一个 float 数组,该数组包含 Color 的颜色分量和 alpha 分量
getTransparency()	int	返回此 Color 的透明模式
hashCode()	int	计算此 Color 的哈希码
HSBtoRGB(float hue,float saturation,float brightness)	static int	将由 HSB 模型指定的颜色分量转换为等价的默认 RGB 模型的值的集合
RGBtoHSB(int r,int g,int b,float[] hsbvals)	static float[]	将默认 RGB 模式指定的颜色分量转换为等价的色调、饱和度和亮度值的集合,这三个值是 HSB 模型的三个分量
toString()	String	返回此 Color 的字符串表示形式

第 2 章 标识符、关键字和数据类型

2.1 从下列字符串中选出正确的 Java 关键字。

abstract, bit, boolean, case, character, comment, double, else, end, endif, extend, false, final, finally, float, for, generic, goto, if, implements, import, inner, instanceof, interface, line, long, loop, native, new, null, old, oper, outer, package, print, private, rest, return, short, static, super, switch, synchronized, this, throw, throws, transient, var, void, volatile, where, write

解：Java 关键字包括：

abstract, boolean, case, double, else, false, final, finally, float, for, generic, goto, if, implements, import, inner, instanceof, interface, long, native, new, null, outer, package, private, rest, return, short, static, super, switch, synchronized, this, throw, throws, transient, void, volatile。

不是 Java 关键字的有：

bit, character, comment, end, endif, extend, line, loop, old, oper, print, var, where, write。

2.2 请写出几个正确的 Java 标识符，并试着将它们用到自己设计的程序中。例如用变量 IntegerValue 表示一个整型量，用 MyTestClass 表示自己定义的一个类等。

解：定义一个雇员 Employee 类，它的每个实例用来记录每名雇员的基本情况，包括姓名 name、年薪 salary 及受雇时间 hireDay。这 3 个量都是私有变量，其中，姓名是字符串型，年薪是双精度型。定义一个 MyDate 类（日期类）作为受雇时间，其中含有雇员参加工作的年、月、日。本例中，定义了以下的标识符，其中用作变量名的是 year、month、day、staff、byPercent、name、salary、hireDay；用作方法名的是 getYear、getMonth、getDay、print、raiseSalary、hireYear；用作类名的是 MyDate、Employee。最后，用 MyTestClass 类对该类进行测试，同时用 integerValue 变量表示临时值，用 y、m、d、n、s 表示方法中的形式参数。

将这些标识符用于下面的程序中：

```
import java.util.*;

class MyDate                                //定义自己的日期类
{   private int year;
    private int month;
    private int day;
```

```java
    public MyDate(int y, int m, int d)            //构造方法
    {   year=y;
        month=m;
        day=d;
    }
    public MyDate()                                //构造方法
    {   year=2012;
        month=11;
        day=8;
    }
    public int getYear()                           //返回年
    {   return year;
    }
    public int getMonth()                          //返回月
    {   return month;
    }
    public int getDay()                            //返回日
    {   return day;
    }
    public void setDate(MyDate SpeDate)            //设置日期
    {   year=SpeDate.getYear();
        month=SpeDate.getMonth();
        day=SpeDate.getDay();
    }
}
class Employee
{   private String name;                           //定义雇员类,包括姓名、年薪及参加工作日期
    private double salary;
    private MyDate hireDay;

    public Employee(String n, double s, MyDate d)
    {   name=n;
        salary=s;
        hireDay=d;
    }
    public void print()
    {   System.out.println(name+" "+salary+" "+hireYear());
    }
    public void raiseSalary(double byPercent)
    {   salary *=1+byPercent /100;
    }
    public int hireYear()
```

```
    {   return hireDay.getYear();
    }
}

public class MyTestClass
{   public static void main(String[] args)
    {   Employee[] staff=new Employee[3];

        staff[0]=new Employee("Harry Hacker", 35000, new MyDate(1989,10,1));
        staff[1]=new Employee("Carl Cracker", 75000, new MyDate(1987,12,15));
        staff[2]=new Employee("Tony Tester", 38000, new MyDate(1990,3,15));

        int integerValue;
        System.out.println("The information of Employee are:");
        for (integerValue=0; integerValue<3; integerValue++)
            staff[integerValue].raiseSalary(5);
        for (integerValue=0; integerValue<3; integerValue++)
            staff[integerValue].print();
    }
}
```

程序中已经输入了 3 名雇员的基本信息,并按 5% 的比例涨年薪。使用命令行编译并运行该程序,程序的执行结果如图 2-1 所示。

图 2-1 MyTestClass 的执行结果

程序中给出的 setDate() 方法并没有在测试类 MyTestClass 中调用,可以使用下列语句进行测试:

```
MyDate birthday=new MyDate();
MyDate yourbirthday=new MyDate();
yourbirthday.setDate(birthday);
```

使用下列语句看看执行的结果:

```
System.out.println("birthday:"+birthday.getYear()+"/"+birthday.getMonth()
+"/"+birthday.getDay());
System.out.println("yourbirthday:"+yourbirthday.getYear()+"/"+
yourbirthday.getMonth()+"/"+yourbirthday.getDay());
```

2.3 请叙述标识符的定义规则。指出在下面的标识符中，哪些是不正确的，并说明原因。

here,there,this,that,it,2to1

解：变量、常量、方法、对象和类等需要使用一个名字来表示，这个名字就是标识符。就是说，在程序中标识符可用作变量名、方法名、接口名、类名等。一般而言，标识符由编程者自行定义，但要遵循相应的语法规则。在 Java 中，对于标识符的规则主要有：

(1) 标识符是以字母、下画线(_)或美元符($)开头，由字母、数字、下画线(_)或美元符号($)组成的字符串。

(2) 标识符区分大小写。

(3) 标识符的长度没有限制。

(4) 注释不能插在一个标识符或关键字之中。

(5) Java 有许多关键字，它们都有各自的特殊意义和用法，不得用它们作为标识符。标识符内可以包含关键字，但不能与关键字完全一样。如 thisOne 是一个合法的标识符，但 this 是关键字，不能当作标识符。

除以上所列几项之外，标识符中不能含有其他符号，例如＋、＝、、*及％等，当然也不允许插入空格。

题目所给的标识符中，不正确的有 this、2to1，其中 this 是 Java 中的关键字，不能用作标识符，2to1 是以数字开头的字符串，也不符合 Java 对标识符的规定。其他的都是正确的标识符。

【拓展思考】

(1) 为什么下列有效的 Java 标识符不是好的标识符？

q,totVal,theNextValueInTheList

解：这 3 个字符串都是合法有效的标识符，它们完全符合标识符的定义规则。命名标识符时，除了要求必须符合规则外，还应尽量遵从一些命名惯例。包括：尽量不使用单字符标识符，除非用于循环控制变量；不使用无意义的或是含义不清的缩写；不使用冗长的标识符；不要使用只是大小写有区别的多个标识符等。

q 是单字符标识符，totVal 的含义并不清楚，而 theNextValueInTheList 太冗长了。

(2) Java 是大小写敏感的。这代表什么意思？

解：大小写字母都可以用在标识符中，而且字母大小写是区分的。Java 对字母大小写是敏感的，意味着只是字母的大小写不同的两个标识符被看作是不同的标识符。所以，total、Total、ToTaL 和 TOTAL 都是不同的标识符。

(3) 什么是空白？它对程序的运行有何影响？它对程序的可读性有何影响？

解：空白是指空格、制表符和换行符，它们用来分隔程序中的字和符号。编译程序忽略额外的空白，所以它不影响执行。适当地使用空白，让程序易读是很重要的。

(4) 下列符号中哪些不是有效的 Java 标识符？为什么？

① RESULT

② result

③ 12345

④ x12345y

⑤ black&white

⑥ answer_7

解：除 12345（因为标识符不能以数字开头）和 black&white（标识符中不能含有 & 字符）之外，所有的标识符都是有效的。标识符 RESULT 和 result 都是有效的，但不应该同时用在一个程序中，因为它们只是大小写不同。下画线（answer_7 中所用的）可用在标识符中。

2.4 Java 中共有哪些基本数据类型？它们分别用什么符号来表示？各自的取值范围是多大？试对每种数据类型定义一个变量，并给它赋一个值。

解：基本数据类型是如数或字符这样的基本值。Java 中定义的各种数据类型占用固定的内存长度，与当前系统的软硬件环境无关。例如，int 型肯定是 32 位的整数，而不管是在 Unix、Windows 还是 Macintosh 环境下。这个特点体现了 Java 的平台无关性。另外 Java 还为每种数据类型都预定义了一个默认值，以保证在任何情况下对变量的取值都是正确的。这个特点体现了 Java 的安全稳定性。

Java 的数据类型共分为两大类，一类是基本类型，一类是复合数据类型。基本类型共有 8 种，分为 4 小类，分别是逻辑型、字符型、整型和浮点型。

逻辑类型或称布尔类型使用 boolean 表示，它有两个常量值：true 和 false，它们全使用小写字母。

字符类型使用 char 表示，一个 char 表示一个 Unicode 字符。每一个 Unicode 字符可用'\uxxxx'表示，其中 xxxx 是任意的一个 16 位无符号整数，范围为 0~65 535。char 类型的常量值必须用一对单引号('')括起来。例如，'B'的 Unicode 值为'\u0042'。

整数类型共有 4 种，分别是：byte(1 字节)、short(2 字节)、int(4 字节)及 long(8 字节)。byte 表示的数的范围为 -128~127，short 表示的数的范围为 -32 768~32 767，int 表示的数的范围为 -2 147 483 648~2 147 483 647，long 表示的数的范围为 -9 223 372 036 854 775 808~9 223 372 036 854 775 807。整型常量可用十进制、八进制或十六进制形式表示。以 1~9 开头的数为十进制数，以 0 开头的数为八进制数，以 0x 开头的数为十六进制数。Java 中 4 种整型量都是有符号的。

浮点数共有两种类型，分别是：单精度浮点数 float(4 字节)及双精度浮点数 double(8 字节)。float 表示的数的范围约为 $1.4e^{-45} \sim 3.4e^{+38}$，double 表示的数的范围约为 $4.9e^{-324} \sim 1.8e^{+308}$。一般地，为了区分两种浮点数常量，通常在常量后加一个字符，加 f 表示是单精度型浮点数，加 d 表示是双精度型浮点数。

下面的这个实例中为每种基本类型定义了一个变量，并为其赋值。

程序代码实现如下：

```
import java.util.*;

public class DataType
{   public static void main(String[] args)
```

```
{   boolean  flag;
    char     yesChar;
    byte     finByte;
    int      intValue;
    long     longValue;
    short    shortValue;
    float    floatValue;
    double   doubleValue;

    flag        =true;
    yesChar     ='y';
    finByte     =30;
    intValue    =-70000;
    longValue   =2001;                          //使用 l 表示是长整型常量
    shortValue  =20000;
    floatValue  =9.997E-5f;                     //使用 f 表示是浮点型常量
    doubleValue=floatValue*floatValue;

    System.out.println("The values are:");
    System.out.println("布尔类型变量         flag="+flag);
    System.out.println("字符类型变量         yesChar="+yesChar);
    System.out.println("字节类型变量         finByte="+finByte);
    System.out.println("整型变量             intValue="+intValue);
    System.out.println("长整型变量           longValue="+longValue);
    System.out.println("短整型变量           shortValue="+shortValue);
    System.out.println("浮点型变量           floatValue="+floatValue);
    System.out.println("双精度浮点型变量 doubleValue="+doubleValue);
    }
}
```

程序运行结果如图 2-2 所示。

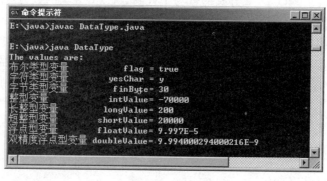

图 2-2　DataType 的运行结果

2.5　什么是对象？基本数据类型与对象有何不同？

解：对象有两个层次的含义：现实生活中对象指的是客观世界的实体；在程序中对象就是一组变量和相关方法的集合，其中变量表明对象的状态，方法表明对象所具有的行为。

对象是类的一个实例。类可以看作是一个模板——正在描述的对象的模型。每当创建一个类的实例时，得到的就是一个对象。

说明一个基本类型的变量时，它可以是 boolean、byte、short、char、int、long、float 或 double 类型中的一种，相应地系统要为它分配内存空间。

使用类类型说明变量。如 String 或用户定义的任何类型，系统都不自动分配内存空间，需要由程序员使用语句完成。实际上，Java 对类类型变量的内存分配分两步进行。说明变量时，在内存中先为其建立一个引用，并置初值 null，表示不指向任何内存空间。然后，由程序员通过 new 语句申请相应的内存空间，内存空间的大小依 class 的定义而定，并将该段内存的首地址赋给刚才建立的引用。换句话说，用 class 类型说明的一个变量并不是数据本身，而只是对数据的引用，进一步要用 new 来创建类的实例或称为对象。

基本数据类型是如数或是字符这样的基本值。对象是更复杂的项，通常包含了定义对象的基本数据类型。它们所对应的内存分配的时机、使用的方式都是不同的。

2.6 Java 中的类型转换是什么？如何进行安全的类型转换？

解：Java 中的类型转换是将不同类型的数据转换为同一类型的数据。

转换的一般原则是将位数少的类型转换为位数多的类型，这称作自动类型转换，也称为加宽转换。相反的转换称为缩窄转换，即占用空间位数较多的数据值转换为占用空间位数较少的数据值。

加宽转换可由系统自动完成，这样规定可以保证转换时不丢失有用信息。而在缩窄转换中，信息很可能会丢失，故缩窄转换不如加宽转换安全。

各类型所占用的位数从少到多依次为：

byte short char int long float double

因此它们之间进行自动类型转换时，原则上是排在前面的类型向排在后面的类型转换，我们将具体的转换规则列在表 2-1 中。

表 2-1 不同类型数据的转换规则

操作数 1 类型	操作数 2 类型	转换后的类型
byte 或 short	int	int
byte 或 short 或 int	long	long
byte 或 short 或 int 或 long	float	float
byte 或 short 或 int 或 long 或 float	double	double
char	int	int

虽然 int 型与 float 型占用的位数一样多，long 型与 double 型占用的位数一样多，但由于浮点数表示的数的范围远远大于整型数，所以，由 int 和 long 型向 float 或 double 型

的转换属于加宽转换,且是自动进行的。但是转换过程中,可能会损失数的精度。使用下面这个程序进行说明。结果如图 2-3 所示。

```java
import java.util.*;

public class DataType
{   public static void main(String[] args)
    {   int       intValue;
        long      longValue;
        float     floatValue;
        double    doubleValue;

        intValue    =1234567892;
        longValue   =1234567890123456789L;

        floatValue  =intValue;
        doubleValue =intValue;
        System.out.println("整型变量            intValue="+intValue);
        System.out.println("浮点型变量          floatValue="+floatValue);
        System.out.println("双精度浮点型变量 doubleValue="+doubleValue);


        floatValue=longValue;
        doubleValue=longValue;
        System.out.println("长整型变量          longValue="+longValue);
        System.out.println("浮点型变量          floatValue="+floatValue);
        System.out.println("双精度浮点型变量 doubleValue="+doubleValue);
    }
}
```

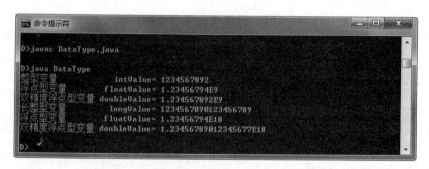

图 2-3 类型转换示例

当位数多的类型向位数少的类型进行转换时,需要用户明确指明,这种转换称为强制类型转换。例如:

```
int i=3;
byte b=(byte) i;
```

将 int 型变量 i 赋给 byte 型变量 b 之前,先将 i 强制转为 byte 型,然后再赋值。一般地,高级类型(即位数较多的数据类型)转为低级类型(即位数较少的数据类型)时,截断高位内容,因此会导致精度下降或数据溢出。

2.7 什么是类?什么是面向对象的程序设计方法?你学过哪些程序设计方法?试着比较这些设计方法的异同点。

解:面向对象的程序设计方法或面向对象编程简称为 OOP,即"Object-Oriented Programming"。这是目前占据主流地位的一种程序设计方法,取代了以前所谓的结构化的以过程为基础的程序设计方法。Java 是一种完全面向对象的语言,使用它可以进行面向对象的程序设计。

在面向对象程序设计方法(OOP)出现之前,软件界广泛流行的是面向过程的设计方式。这种方法处理的重点集中在函数或是方法,程序中的任何地方都可以通过名字并使用参数调用方法。同时程序中使用的众多变量名、函数名互不约束,令程序员不堪重负。特别是当开发大型系统时,需要多人合作完成开发项目,每个人负责自己的一部分工作,如果想读懂合作者的代码简直是不可能的。这种情况在程序维护时矛盾突出,问题很多。另一方面,每个人的代码都可能被其他人所用,读不懂的代码当然谈不上很好利用。例如,由不同程序员分别负责的对内存空间的申请与释放就难于控制,因此经常会出现所谓的内存漏洞问题。由于使用面向过程方法设计的程序把处理的主体与处理的方法分开,因此各种成分错综复杂地放在一起,难以理解,易出错,并且难于调试。这样的设计方法往往限制了开发项目的规模。

随着开发系统的不断增大、复杂,面向过程的方法越来越不能满足开发者的要求,面向对象的技术应运而生。OOP 技术使得程序结构简单,相互协作容易,更重要的是程序的重用性大大提高了。

OOP 技术把问题看成是相互作用的事物的集合,用属性来描述事物,而把对它的操作定义为方法。在 OOP 中,把事物称为对象,把属性称为数据,那么对象就是数据加方法。

OOP 中采用了三大技术:封装、继承和多态。将数据及对数据的操作捆绑在一起成为类,这就是封装技术。程序员只有一种基本的结构,即类。将一个已有类中的数据和方法保留,并加上自己特殊的数据和方法,从而构成一个新类,这就是 OOP 中的继承。原来的类是父类,新类是子类,父类派生了子类,或说子类继承于父类。在一个类或多个类中,可以让多个方法使用同一个名字,从而具有多态性。多态可以保证对不同类型的数据进行等同的操作,名字空间也更加宽松。

使用面向过程编程时,程序员通过步步细化的过程,将准备完成的任务分解成几个子任务,然后再分解成更小的子任务,直到子任务简单到足以直接编程为止。这是自顶向下的编程方式。与此相对应的是自底向上的编程方式,即先编写解决简单任务的过程,并把它们组合成更大更复杂的过程,直到它们的功能满足系统需求为止。这两种方法都是面向过程编程中经常使用的方法。在实际实现过程中,程序员往往将两者相结合,灵活利用自顶向下及自底向上的方法。

在实际的开发过程中,面向对象的设计方法和面向过程的设计方法往往相辅相成。

在系统级设计时,经常使用面向对象的设计方法,定义要使用的类及其相关的成员变量和成员方法,定义所需的接口。而在方法的具体实现时,多采用面向过程的设计方法。就一个函数或是过程的内部实现来看,面向过程仍然是一种不可缺少的设计方式。

【拓展思考】
支持面向对象程序设计的基本概念是什么?

解:支持面向对象程序设计的基本元素是对象、类、封装和继承。对象由类来定义,类包括了对那些对象定义操作的方法(它们执行的服务)。对象是封装的,这样它们仅保存及管理自己的数据。继承是一个类由另一个类派生的复用技术。

2.8 学习过 C++ 的读者,可以比较 Java 与 C++ 在面向对象设计方面的不同点和相同点。

解:Java 与 C++ 有很多相同的地方。它们都有类的概念,类中的基本内容也大同小异。

它们都采用了 OOP 中的三大技术,即封装、继承和多态。除此之外,它们也存在某些差别,其中最大的一点是,C++ 有多重继承能力。多重继承是指从多个类派生一个子类,即一个类可以有多个父类。多重继承关系类似于一个网。如图 2-4 所示,父类 1、父类 2 及父类 3 共同派生子类 1,而父类 2 和父类 4 派生子类 2。

由于多个父类派生一个子类,因此子类可以继承这若干个父类的特性。这虽然使得特性继承比较灵活,但也会造成混乱。具体来说,如果子类的多个父类中有同名的方法和数据,那么容易造成子类实例的混乱。如上例中,如果子类 1 的 3 个父类中都有一个 MinParent() 方法,方法的参数列表是完全一样的,而子类 1 又没有重写这个方法,当子类 1 的实例需要调用这个方法时,系统将不知道该调用哪个父类中的同名方法,这是多重继承不可克服的缺点。Java 抛弃了多重继承,只允许单重继承,即如果继承的话,只能有一个父类。单重继承的关系类似于一棵树,子类与父类之间的关系非常清楚,不会造成任何混乱。图 2-5 所示的是一个单重继承的例子,图中每个子类都只有一个父类。

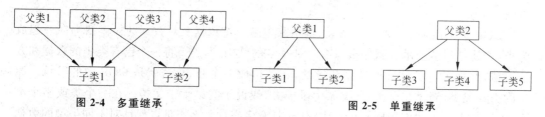

图 2-4 多重继承　　　　　　　　　　图 2-5 单重继承

虽然 Java 中去除了多重继承的写法,但并没有减弱这方面的能力。Java 中提供了接口这个新概念。接口是一种特殊的类,多重继承的能力通过接口来实现。当一个子类需要多重继承时,可以将被继承的多个类设计为接口,这是特殊的"父类"。类中的方法都不含有方法体,方法体均在子类中具体实现。即使多个"父类"中有同名的方法,由于方法体只在子类中出现,所以也不会出现混乱,对应的代码是唯一的。

Java 在类层次之上又提出了包的概念,可减少命名冲突,扩大名字空间。

2.9 试编码定义一个公有类 PubTest1,它含有两个浮点类型变量 fvar1 和 fvar2,还有一个整数类型的变量 ivar1。PubTest1 类中有一个方法 sum_f_I(),它将 fvar1 与 ivar1

的值相加,结果放在 fvar2 中。

解:定义公有类 PubTest1,它的 3 个成员变量分别是两个浮点类型变量 fvar1、fvar2 和一个整数类型变量 ivar1。其中,fvar1 和 ivar1 存放程序中给定的初值,fvar2 用来存放前两个变量计算的结果。生成 PubTest1 的一个实例时,只给 fvar2 赋初值 0。注意此处使用的是浮点类型的常量,但没有给另外的两个量赋值。为此设计两个成员方法 setFvar1() 和 setIvar1(),分别给 fvar1 和 ivar1 赋值。print() 方法则将计算的结果显示出来。

程序代码实现如下:

```java
import java.util.*;

class PubTest1                                    //定义公有类
{    private int ivar1;                           //类 PubTest1 中的 3 个数据成员
     private float fvar1,fvar2;

     public PubTest1()
     {   fvar2=0.0f;                              //只给 fvar2 赋初值
     }
     public float sum_f_I ()                      //计算 fvar2 的值
     {   fvar2=fvar1+ivar1;
         return fvar2;
     }
     public void print()                          //显示 fvar2 的值
     {   System.out.println("fvar2="+fvar2);
     }
     public void setIvar1(int ivalue)             //给 ivar1 赋值
     {   ivar1=ivalue;
     }
     public void setFvar1(float fvalue)           //给 fvar1 赋值
     {   fvar1=fvalue;
     }
}
public class PubMainTest
{    public static void main(String[] args)
     {   PubTest1 pubt1=new PubTest1();

         pubt1.setIvar1(10);                      //给 pubt1 对象的 ivar1 赋值
         pubt1.setFvar1(100.02f);                 //给 pubt1 对象的 fvar1 赋值
         pubt1.sum_f_I();                         //计算 fvar2 的值
         pubt1.print();                           //输出 fvar2 的值
     }
}
```

程序的执行结果如图 2-6 所示。

图 2-6　PubTest1 的执行结果

2.10 以一个高等学校中的各类人员为研究对象，例如可以定义教师类，包括以下属性：姓名、性别、出生日期、工资号、所在系所、职称等基本信息，同时还可以定义相关的方法。教师类又可以包含研究系列、实验系列、图书管理系列、行政系列等子类，这些子类可以从父类中继承属性和方法，也可以再定义其他的属性和方法。请参照这个例子，定义一所高校中包含的各类人员的类（要求至少定义 5 个类），为每个类指明它应有的属性，并按实际情况组织类的层次。对每个类的每个属性，要定义必要的访问方法。

解：根据题意，可以定义一个教师类，它包括 4 个子系列的教师，分别是：研究系列、实验系列、图书管理系列、行政系列。所有教师都有一些共同的属性，各个子系列的教师又具有区别于其他系列的特殊属性。因此我们定义一个基类 Teacher，它包括的通用属性有：姓名、性别、出生日期、工资号、所在系所、职称等；基类下派生 4 个子类，分别是：研究系列教师 ResearchTeacher、实验系列教师 LabTeacher、图书管理系列教师 LibTeacher 和行政系列教师 AdminTeacher。这些子类除继承父类的属性和方法外，还具有各自独特的属性和方法。其中为研究系列子类定义了一个特殊的属性即研究领域 resField，为实验系列子类定义了一个特殊属性即所属实验室名称 labName，为图书管理系列子类定义一个特殊属性即所在的图书馆阅览室名 libName，为行政系列子类定义一个特殊属性即其管理职务 managePos。针对这些属性，定义了各自相应的访问方法。对于日期属性，类似于习题 2.2 中的处理，定义了日期类 Date。

程序代码实现如下：

```java
import java.lang.*;

class Date                                    //定义日期类
{   int day;
    int month;
    int year;
    Date (int day, int month, int year)
    {   this.day=day;
        this.month=month;
        this.year=year;
    }
    Date ()
    {   this.day=8;
        this.month=11;
```

```java
        this.year=2012;
    }
    public int getYear()                        //返回年
    {   return year;
    }
    public int getMonth()                       //返回月
    {   return month;
    }
    public int getDay()                         //返回日
    {   return day;
    }
    public void setDate(Date SpeDate)           //设置日期
    {   year=SpeDate.getYear();
        month=SpeDate.getMonth();
        day=SpeDate.getDay();
    }
}

public class Teacher                            //定义教师类,这是父类
{   protected String name;                      //教师名字
    protected boolean sex;                      //性别,true 表示男性,false 表示女性
    protected Date birth;                       //教师出生日期
    protected String salaryID;                  //教师工资号
    protected String depart;                    //教师所属系
    protected String posit;                     //教师职称

    String getName()                            //返回教师名字
    {   return name;
    }
    void setName (String name)                  //记录教师名字
    {   this.name=name;
    }
    boolean getSex()                            //返回教师性别
    {   return sex;
    }
    void setSex (boolean sex)                   //记录教师性别
    {   this.sex=sex;
    }
    Date getBirth()                             //返回教师出生日期
    {   return birth;
    }
    void setBirth (Date birth)                  //记录教师出生日期
    {   this.birth=birth;
    }
```

```java
    String getSalaryID()                            //返回教师工资号
    {   return salaryID;
    }
    void setSalaryID (String salaryID)              //记录教师工资号
    {   this.salaryID=salaryID;
    }
    String getDepart()                              //返回教师所属系
    {   return depart;
    }
    void setDepart (String depart)                  //记录教师所属系
    {   this.depart=depart;
    }
    String getPosit()                               //返回教师职称
    {   return posit;
    }
    void setPosit (String posit)                    //记录教师职称
    {   this.posit=posit;
    }

    public Teacher(String name, boolean sex, Date birth,
            String salaryid, String depart, String posit)
    {   this.name=name;                             //构造函数
        this.sex=sex;
        this.birth=birth;
        this.salaryID=salaryid;
        this.depart=depart;
        this.posit=posit;
    }
    public Teacher()
    {   this.name="Virtual man";                    //构造函数
        this.sex=true;
        this.birth=new Date();
        this.salaryID="007";
        this.depart="keep secret";
        this.posit="unknown";
    }
    public void print()                             //输出教师基本信息
    {   System.out.print("The teacher name:  ");
        System.out.println(this.getName());
        System.out.print("The teacher sex:  ");
        if (this.getSex()==false)
        {   System.out.println("女");
        }
        else
```

```java
        {   System.out.println("男");
        }
        System.out.print("The teacher birth:   ");
        System.out.println(this.getBirth().year +"-"+this.getBirth().month
            +"-"+this.getBirth().day );
        System.out.print("The teacher salaryid:   ");
        System.out.println(this.getSalaryID());
        System.out.print("The teacher posit:   ");
        System.out.println(this.getPosit());
        System.out.print("The teacher depart:   ");
        System.out.println(this.getDepart());
    }
    public static void main (String [] args)
    {   Date dt1=new Date(12, 2, 1985);              //创建日期实例,作为教师出生日期
        Date dt2=new Date(2, 6, 1975);
        Date dt3=new Date(11, 8, 1964);
        Date dt4=new Date(10, 4, 1975);
        Date dt5=new Date(8, 9, 1969);

        //创建各系列教师实例,用来进行测试
        Teacher t1=new Teacher("zhangsan", false,dt1, "123", "CS", "Prefessor");
        ResearchTeacher rt=new ResearchTeacher("lisi", true, dt2, "421",
            "software engineering", "associate professor", "Software");
        LabTeacher lat=new LabTeacher("wangwu", false, dt3, "163",
            "Foreign Language", "instructor", "Speech Lab");
        LibTeacher lit=new LibTeacher("zhouliu", true, dt4, "521",
            "Physics ", "Prefessor", "PhysicalLib");
        AdminTeacher at=new AdminTeacher("zhaoyun", false, dt5, "663",
            "Environment", "Prefessor", "dean");

        //分别调用各自的输出方法,输出相应的信息
        System.out.println("--------------------------------");
        t1.print();                              //输出普通教师的信息
        System.out.println("--------------------------------");
        rt.print();                              //输出研究系列教师的信息
        System.out.println("--------------------------------");
        lat.print();                             //输出实验系列教师的信息
        System.out.println("--------------------------------");
        lit.print();                             //输出图书管理系列教师的信息
        System.out.println("--------------------------------");
        at.print();                              //输出行政系列教师的信息
        System.out.println("--------------------------------");
    }
}
```

```java
class ResearchTeacher extends Teacher                //研究系列教师类的定义
{   private String resField;                         //增加的研究领域属性
    public ResearchTeacher(String name, boolean sex, Date birth,
            String salaryid, String depart, String posit, String resField)
    {   this.name=name;                              //与父类共有的属性的赋值
        this.sex=sex;
        this.birth=birth;
        this.salaryID=salaryid;
        this.depart=depart;
        this.posit=posit;
        this.resField=resField;                      //特殊属性的赋值
    }
    String getResField()                             //特殊属性的访问方法
    {   return resField;
    }
    void setResField (String resField)               //特殊属性的访问方法
    {   this.resField=resField;
    }
    public void print()
    {   System.out.println("One of Research teachers' info is ");
        System.out.print("The teacher name:  ");    //与父类共有的属性的输出
        System.out.println(this.getName());
        System.out.print("The teacher sex:  ");
        if (this.getSex()==false)
        {   System.out.println("女");
        }
        else
        {   System.out.println("男");
        }
        System.out.print("The teacher birth:  ");
        System.out.println(this.getBirth().year +"-"+this.getBirth().month
            +"-"+this.getBirth().day );
        System.out.print("The teacher salaryid:  ");
        System.out.println(this.getSalaryID());
        System.out.print("The teacher posit:  ");
        System.out.println(this.getPosit());
        System.out.print("The teacher depart:  ");
        System.out.println(this.getDepart());
        System.out.print("The teacher research field is:  ");  //特殊属性的输出
        System.out.println(this.getResField());               //特殊属性的输出
    }
}
```

```java
class LabTeacher extends Teacher                    //实验系列老师类的定义
{   private String labName;                         //增加的实验室名称属性
    public LabTeacher(String name, boolean sex, Date birth,
        String salaryid, String depart, String posit, String labName )
    {   this.name=name;                             //与父类共有的属性的赋值
        this.sex=sex;
        this.birth=birth;
        this.salaryID=salaryid;
        this.depart=depart;
        this.posit=posit;
        this.labName=labName;                       //特殊属性的赋值
    }
    String getLabName()                             //特殊属性的访问方法
    {   return labName;
    }
    void setLabName (String labName)                //特殊属性的访问方法
    {   this.labName=labName;
    }

    public void print()
    {   System.out.println("One of Lab teachers' info is ");
        System.out.print("The teacher name:   ");   //与父类共有的属性的输出
        System.out.println(this.getName());
        System.out.print("The teacher sex:   ");
        if (this.getSex()==false)
        {   System.out.println("女");
        }
        else
        {   System.out.println("男");
        }
        System.out.print("The teacher birth:   ");
        System.out.println(this.getBirth().year +"-"+this.getBirth().month
            +"-"+this.getBirth().day );
        System.out.print("The teacher salaryid:   ");
        System.out.println(this.getSalaryID());
        System.out.print("The teacher posit:   ");
        System.out.println(this.getPosit());
        System.out.print("The teacher depart:   ");
        System.out.println(this.getDepart());
        System.out.print("The teacher Lab name is:   ");  //特殊属性的输出
        System.out.println(this.getLabName());            //特殊属性的输出
    }
}
```

```java
class LibTeacher extends Teacher                    //图书管理系列老师类的定义
{   private String libName;                         //增加的图书室名称属性
    public LibTeacher(String name, boolean sex, Date birth,
          String salaryid, String depart, String posit, String libName )
    {   this.name=name;                             //与父类共有的属性的赋值
        this.sex=sex;
        this.birth=birth;
        this.salaryID=salaryid;
        this.depart=depart;
        this.posit=posit;
        this.libName=libName;                       //特殊属性的赋值
    }
    String getLibName()                             //特殊属性的访问方法
    {   return libName;
    }
    void setLibName (String libName)                //特殊属性的访问方法
    {   this.libName=libName;
    }
    public void print()
    {   System.out.println("One of Library teachers' info is ");
        System.out.print("The teacher name:   ");   //与父类共有的属性的输出
        System.out.println(this.getName());
        System.out.print("The teacher sex:   ");
        if (this.getSex()==false)
        {   System.out.println("女");
        }
        else
        {   System.out.println("男");
        }
        System.out.print("The teacher birth:   ");
        System.out.println(this.getBirth().year +"-"+this.getBirth().month
            +"-"+this.getBirth().day );
        System.out.print("The teacher salaryid:   ");
        System.out.println(this.getSalaryID());
        System.out.print("The teacher posit:   ");
        System.out.println(this.getPosit());
        System.out.print("The teacher depart:   ");
        System.out.println(this.getDepart());
        System.out.print("The teacher library field is:   ");   //特殊属性的输出
        System.out.println(this.getLabName());
    }
}

class AdminTeacher extends Teacher                  //行政系列老师类的定义
```

```java
{   String managePos;                                  //增加的管理职位属性
    public AdminTeacher(String name, boolean sex, Date birth,
        String salaryid, String depart, String posit, String managePos )
    {   this.name=name;                                //与父类共有的属性的赋值
        this.sex=sex;
        this.birth=birth;
        this.salaryID=salaryid;
        this.depart=depart;
        this.posit=posit;
        this.managePos=managePos;                      //特殊属性的赋值
    }
    String getManagePos()                              //特殊属性的访问方法
    {   return managePos;
    }
    void setManagePos(String managePos)                //特殊属性的访问方法
    {   this.managePos=managePos;
    }
    public void print()
    {   System.out.println("One of Administratary teachers' info is ");
        System.out.print("The teacher name:  ");       //与父类共有的属性的输出
        System.out.println(this.getName());
        System.out.print("The teacher sex:  ");
        if (this.getSex()==false)
        {   System.out.println("女");
        }
        else
        {   System.out.println("男");
        }
        System.out.print("The teacher birth:  ");
        System.out.println(this.getBirth().year +"-"+this.getBirth().month
            +"-"+this.getBirth().day );
        System.out.print("The teacher salaryid:  ");
        System.out.println(this.getSalaryID());
        System.out.print("The teacher posit:  ");
        System.out.println(this.getPosit());
        System.out.print("The teacher depart:  ");
        System.out.println(this.getDepart());
        System.out.print("The teacher Management position is:  ");
                                                       //特殊属性的输出
        System.out.println(this.getManagePos());
    }
}
```

程序运行结果如图 2-7 所示。

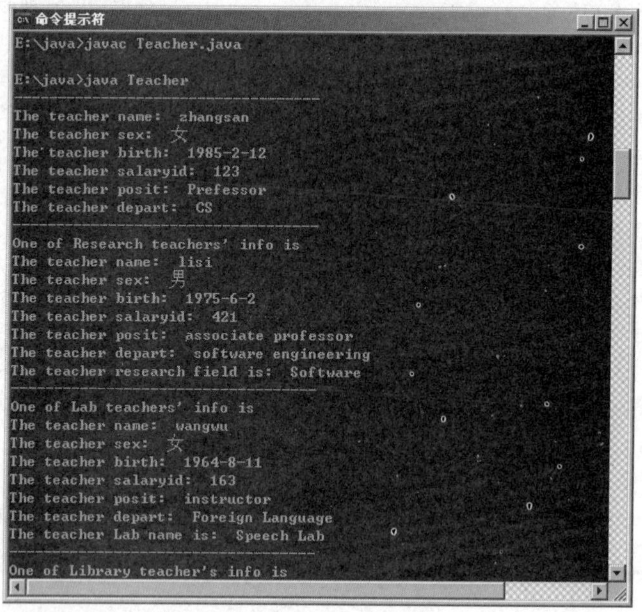

图 2-7 Teacher 类运行结果

父类中的一些属性由于也可以由子类来使用,所以类型说明为 protected 型的。这意味着父类中的属性可以由其子类来访问。

从本例的程序实现中可以看出,父类和子类有很多共享的属性,因此它们各自的构造方法很类似。对于输出方法 print()也是这样,也有很多类似的代码。

例如,父类 Teacher 与子类 ResearchTeacher 的构造方法分别是:

```
public Teacher(String name, boolean sex, Date birth,
       String salaryid, String depart, String posit)
{   this.name=name;
    this.sex=sex;
    this.birth=birth;
    this.salaryID=salaryid;
    this.depart=depart;
    this.posit=posit;
}
public ResearchTeacher(String name, boolean sex, Date birth,
       String salaryid, String depart, String posit, String resField)
{   this.name=name;                         //与父类共有的属性的赋值
    this.sex=sex;
    this.birth=birth;
    this.salaryID=salaryid;
    this.depart=depart;
    this.posit=posit;
    this.resField=resField;                 //特殊属性的赋值
}
```

实际上,当子类有很多的属性继承于父类时,其子类的构造方法也可以继承父类的构造方法,构造方法之外的方法也可以借助于父类的同名方法。当使用父类中的方法时,使用关键字 super。因此可以进一步修改本题,简化代码。相应的解决方案参见习题 5.7。

2.11 设计并实现一个 Course 类,它代表学校中的一门课程。按照实际情况,将这门课程的相关信息组织成它的属性,并定义必要的访问方法。

解: 一门课程一般包括如下属性:课程代号、课程名称、课程类别、学时、学分等。根据题意,可以定义 Course 类,包括上述各属性,并在类中定义各个属性相关的访问方法,最后使用主函数测试这个类,并修改课程的学分。

程序代码实现如下:

```java
public class Course
{   private String courseID;                    //定义课程代号
    private String courseName;                  //定义课程名称
    private String courseType;                  //定义课程类别
    private int classHour;                      //定义课程学时
    private float credit;                       //定义课程学分

    public Course(String courseID, String courseName,
        String courseType, int classHour, float credit)
    {   this.courseID=courseID;
        this.courseName=courseName;
        this.courseType=courseType;
        this.classHour=classHour;
        this.credit=credit;
    }

    String getID()                              //得到课程代号
    {   return courseID;
    }
    void setID (String id)                      //设置课程代号
    {   this.courseID=id;
    }
    String getName()                            //得到课程名称
    {   return courseName;
    }
    void setName (String name)                  //设置课程名称
    {   this.courseName=name;
    }
    String getType()                            //得到课程类别
    {   return courseType;
    }
    void setType (String courseType)            //设置课程类别
```

```java
        {   this.courseType=courseType;
        }
        int getClassHour()                              //得到课时数
        {   return classHour;
        }
        void setClassHour (int classHour)               //设置课时数
        {   this.classHour=classHour;
        }

        float getCredit()                               //得到课程学分数
        {   return credit;
        }
        void setCredit(float credit)                    //设置课程学分数
        {   this.credit=credit;
        }

        public void print()                             //打印输出课程基本信息
        {   System.out.println("The basic info of this course as followed:");
            System.out.println("courseID :"+this.getID()) ;
            System.out.println("courseName:"+this.getName());
            System.out.println("courseType:"+this.getType());
            System.out.println("classHour:"+this.getClassHour());
            System.out.println("credit:"+this.getCredit());
            System.out.println("--------------------------------");
        }

        public static void main(String[] args)
        //测试课程类 Course 的定义,修改课程学分
        {   //创建一个课程实例,最后一项的输入是 3.0f,表示是浮点型常量
            Course cs=new Course("D13", "Java Program Design", "CS", 64, 3.0f);
            System.out.println("--------------------------------");
            cs.print();
            System.out.println("Alter the credit as a new value 4.0.");
            cs.setCredit(4.0f);                         //修改课程学分,由 3 分改为 4 分
            cs.print();
        }
    }
```

程序的执行结果如图 2-8 所示。

2.12 设计并实现一个 MyGraphic 类及其子类,它们代表一些基本的图形,这些图形包括矩形、三角形、圆、椭圆、菱形、梯形等。试给出能描述这些图形所必需的属性及必要的方法。

解:根据题意,程序设计了图形类型 MyGraphic,这是一个基类,它包括两个基本属

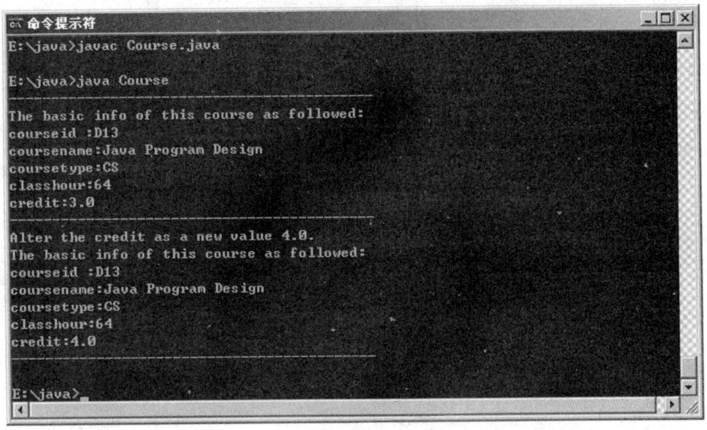

图 2-8 Course 类的执行结果

性,分别是图形线条的颜色 lineColor 和图形的填充颜色 fillColor。作为示例,还设计了矩形类 MyRectangle 和圆类 MyCircle 两个子类,其中矩形类有基本属性长 rLong 和宽 rWidth,使用方法 float calCircum()可以计算矩形的周长,使用方法 float calSquare()可以计算矩形的面积;圆类包括基本属性半径 radius,也可以通过上述两个同名的方法计算其圆周长和面积。其余各图形子类的实现可以参照这两个类的实现来完成。

对于子类继承于父类的属性,在构造方法中借用父类的构造方法对其进行赋值。例如 MyRectangle 类实例的 4 个属性中,其中的线条颜色 lc 和填充颜色 fc 将使用父类的构造方法,如下所示:

```
MyRectangle(float rl, float rw, String lc, String fc)
{   super(lc, fc);              //调用父类的方法,设置线条颜色和填充颜色
    this.rLong=rl;
    this.rWidth=rw;
}
```

程序代码实现如下:

```
public class MyGraphic
{   String lineColor;                           //线条颜色
    String fillColor;                           //填充颜色
    MyGraphic(String lc, String fc)             //图形的构造方法
    {   this.lineColor=lc;
        this.fillColor=fc;
    }
    void print()
    {   System.out.println("line color is "+this.lineColor+
            "\t fill color is "+this.fillColor);
    }

    public static void main(String[] args)
```

```java
        {   //常量4.5默认为是双精度型的,需要进行强制类型转换;4.5f是浮点型常量
            float rd=(float)4.5;
            //定义一个"黑边白心"的圆
            MyCircle mc=new MyCircle(rd, "black", "white");
            //定义一个"红边蓝心"的矩形
            MyRectangle mr=new MyRectangle(4, 6, "red", "blue");
            System.out.println("Circle info:");
            mc.print();
            System.out.println("circumference is "+mc.calCircum());    //计算圆的周长
            System.out.println("square is "+mc.calSquare());           //计算圆的面积
            System.out.println("Rectangle info:");
            mr.print();
            System.out.println("circumference is "+mr.calCircum());    //计算矩形周长
            System.out.println("square is "+mr.calSquare());           //计算矩形面积
        }
    }
    class MyRectangle extends MyGraphic        //从MyGraphic派生MyRectangle类
    {   float rLong;                            //矩形的长
        float rWidth;                           //矩形的宽
        MyRectangle(float rl, float rw, String lc, String fc)
        {   super(lc, fc);                      //调用父类的方法,设置线条颜色并填充颜色
            this.rLong=rl;
            this.rWidth=rw;
        }

        float calCircum()                       //计算矩形的周长
        {   return ((float)((this.rLong+this.rWidth) * 2));
        }
        float calSquare()                       //计算矩形的面积
        {   return ((float)(this.rLong * this.rWidth));
        }
    };
    class MyCircle extends MyGraphic            //从MyGraphic派生MyCircle类
    {   float radius;                           //圆的半径
        MyCircle(float rd, String lc, String fc)
        {   super(lc, fc);                      //调用父类的方法,设置线条颜色和填充颜色
            this.radius=rd;
        }
        float calCircum()                       //计算圆的周长
        {   return ((float)(this.radius * 3.14 * 2));
        }
        float calSquare()                       //计算圆的面积
        {   return ((float)(radius * radius * 3.14));
```

            }
    };

程序的执行结果如图 2-9 所示。

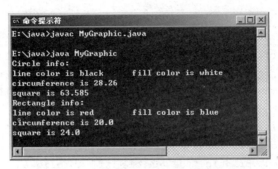

图 2-9　MyGraphic 类的执行结果

2.13　设计并实现一个 Vehicle 类及其子类,它们代表主要的交通工具,定义必要的属性信息及访问方法。

解：根据题意,对于 Vehicle 类定义了基本的属性,包括品牌 brand、颜色 color、价钱 price、数量 number 等。接着定义了该类的两个构造方法,分别用来生成 Vehicle 实例。第一个构造方法带品牌及颜色两个参数,第二个构造方法带全部的 4 个参数。实现时,第二个构造方法可以借用第一个构造方法,完成前两个属性的赋值,如下所示：

```
public Vehicle(String b, String c, int p, int n)
{   this(b,c);                          //调用带两个参数的构造方法
    this.price=p;                       //对价格赋值
    this.number=n;                      //对数量赋值
}
```

在 Vehicle 基础上我们定义了 Car 子类和 Truck 子类,其各自又有新的属性和方法。在这两个子类中,遇到与父类相同的属性时,均调用父类的方法对其进行操作。

调用父类的构造函数的相应语句如下：

```
super(b, c, p, n);                      //使用父类的构造方法
```

调用父类的输出函数的相应语句如下：

```
super.print();                          //调用父类的输出方法
```

注意这两个语句在写法上的差别。

程序代码实现如下：

```
public class Vehicle
{   String brand;                       //品牌
    String color;                       //颜色
    int price;                          //价格
    int number;                         //数量
```

```java
    public Vehicle(String b, String c)
    {   this.brand=b;
        this.color=c;
    }

    public Vehicle(String b, String c, int p, int n)
    {   this(b,c);                              //调用两个参数的构造方法
        this.price=p;                           //对价格进行赋值
        this.number=n;                          //对数量进行赋值
    }
    void print()
    {   System.out.println("\n--------------------------");
        System.out.println("the vehicle info as followed:");
        System.out.print("brand="+this.brand+"\t");
        System.out.print("color="+this.color+"\t");
        System.out.print("price="+this.price+"\t");
        System.out.print("number="+this.number+"\t");
    }
    public static void main(String args[])
    {   Vehicle cl=new Vehicle("vehicle1", "white");
        Vehicle c2=new Vehicle("vehicle2", "white", 300, 1);
        Car cr=new Car("Car1", "red", 300, 4, 400);
        Truck tk2=new Truck("Truck1", "black", 300, 400);

        cl.print();
        c2.print();
        cr.print();
        tk2.print();
    }
};

class Car extends Vehicle                       //Car类
{   int speed;                                  //增加最高时速属性
    Car(String b, String c, int p, int n, int s)
    {   super(b, c, p, n);                      //使用父类的构造方法
        this.speed=s;                           //再对本类的特殊属性赋值
    }
    void print()
    {   super.print();                          //调用父类的输出方法
        System.out.print("speed="+this.speed);  //输出特殊属性
    }
};

class Truck extends Vehicle                     //Truck类
```

```
{   int speed;                                            //最高时速
    int weight;                                           //载重量

    public Truck(String b, String c, int s, int w)
    {   super(b,c);                                       //调用父类的构造方法
        this.speed=s;                                     //对特殊属性进行赋值
        this.weight=w;
    }
    void print()
    {   super.print();                                    //调用父类的输出方法
        System.out.print("speed="+this.speed+"\t");       //输出特殊属性
        System.out.print("weight="+this.weight+"\t");     //输出特殊属性
    }
};
```

程序实现结果如图 2-10 所示。

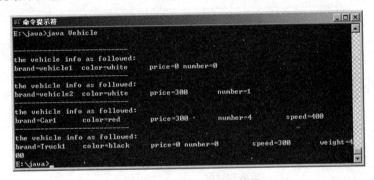

图 2-10　Vehicle 类运行结果

第 3 章　表达式和流程控制语句

3.1 Java 中常用的运算符有哪些？它们的含义分别是什么？

解：Java 运算符按功能可分为：算术运算符、关系运算符、逻辑运算符、位运算符、赋值运算符、条件运算符；除此之外，还有几个特殊用途运算符，如数组下标运算符等。

常用的运算符及其意义如表 3-1 所示。

表 3-1　运算符及其意义

类　型	运　算　符	意　　义
算术运算符	＋	加法或字符串的连接
	－	减法
	＊	乘法
	／	除法，如果参加运算的两个操作数都是整数，则为整除运算，否则为浮点数除法
	％	整数求余运算
	＋＋	加 1 运算
	－－	减 1 运算
	－	求相反数
关系运算符	＝＝	等于
	！＝	不等于
	＞	大于
	＜	小于
	＞＝	大于等于
	＜＝	小于等于
逻辑运算与位运算	＆＆	逻辑与
	‖	逻辑或
	！	取反
	＆	按位与
	｜	按位或
	＾	按位异或
	～	按位取反

续表

类　型	运　算　符	意　　义
逻辑运算与位运算	>>	右移
	>>>	右移,并用0填充高位
	<<	左移
赋值运算符	+=	加法赋值
	-=	减法赋值
	*=	乘法赋值
	/=	除法赋值
	%=	取余赋值
	&=	按位与赋值
	|=	按位或赋值
	^=	按位异或赋值
	<<=	左移赋值
	>>=	右移赋值
	>>>=	右移赋值
其他运算符	[]	数组下标
	.()	方法调用
	?:	三目条件运算
	instanceof	对象运算

3.2　Java中操作符优先级是如何定义的?

解:Java中,各操作符的运算优先顺序如表3-2所示。

表3-2　操作符的运算优先顺序

优　先　级	运　算　符	运　　算	结　合　律
1	[]	数组下标	自左至右
	.	对象成员引用	
	(参数)	参数计算和方法调用	
	++	后缀加	
	--	后缀减	
2	++	前缀加	自右至左
	--	前缀减	
	+	一元加	

续表

优先级	运算符	运算	结合律
2	—	一元减	自右至左
	~	位运算非	
	!	逻辑非	
3	new	对象实例	自右至左
	（类型）	转换	
4	*	乘法	自左至右
	/	除法	
	%	取余	
5	+	加法	自左至右
	+	字符串连接	
	—	减法	
6	<<	左移	自左至右
	>>	用符号位填充的右移	
	>>>	用0填充的右移	
7	<	小于	自左至右
	<=	小于等于	
	>	大于	
	>=	大于等于	
	instanceof	类型比较	
8	==	相等	自左至右
	!=	不等于	
9	&	位运算与	自左至右
	&	布尔与	
10	^	位运算异或	自左至右
	^	布尔异或	
11	|	位或	自左至右
	|	布尔或	
12	&&	逻辑与	自左至右
13	||	逻辑或	自左至右
14	?:	条件运算符	自右至左

续表

优先级	运算符	运算	结合律
15	=	赋值	自右至左
	+=	加法赋值	
	+=	字符串连接赋值	
	-=	减法赋值	
	*=	乘法赋值	
	/=	除法赋值	
	%=	取余赋值	
	<<=	左移赋值	
	>>=	右移(符号位)赋值	
	>>>=	右移(0)赋值	
	&=	位与赋值	
	&=	布尔与赋值	
	^=	位异或赋值	
	^=	布尔异或赋值	
	|=	位或赋值	
	|=	布尔或赋值	

3.3 >>>与>>有什么区别？试分析下列程序段的执行结果：

int b1=1;
int b2=1;

b1 <<= 31;
b2 <<= 31;

b1 >>= 31;
b1 >>= 1;

b2 >>>= 31;
b2 >>>= 1;

解：>>>与>>都是右移运算符，它们的不同之处在于使用不同位填充左侧的空白。>>运算使用符号位填充左侧的空位，而>>>使用零填充空位。也就是说，>>运算保持操作数的符号不变，而>>>运算则可能改变原数的符号。

分析上面的程序段。初始时，b1 和 b2 都是 int 型的变量，占 32 位，初值均为 1，两个变量分别向左移动 31 位，则两个值变为：

```
10000000 00000000 00000000 00000000
```

在计算机内部,这个值是:-2147483648。

下一步,b1 再向右移动 31 位,这里使用的是>>运算符。b1 是一个负数,其最高位为 1,右移时使用 1 填充左侧的空位。右移 31 位后 b1 的值为:

```
11111111 11111111 11111111 11111111
```

在计算机内部,这个值是:-1。b1 继续向右移动 1 位,此时值不变,仍为-1。实际上,此后使用>>运算符将 b1 向右移动任何位,它的值都不会再变了。

使用>>>运算符则有所不同。>>>使用零填充左侧的空位,所以将 b2 向右移动 31 位后,它的值为:

```
00000000 00000000 00000000 00000001
```

即 b2 的值为 1。再向右移动 1 位,则它的值为:

```
00000000 00000000 00000000 00000000
```

即 b2 的值为 0。

测试这些语句的程序如下所示:

```java
import java.util.*;

public class Test
{   public static void main(String[] args)
    {   int b1=1;                              //b1 赋初值
        int b2=1;                              //b2 赋初值
        System.out.println("b1="+b1);
        System.out.println("b2="+b2);

        b1 <<=31;                              //b1 左移 31 位
        b2 <<=31;                              //b2 左移 31 位
        System.out.println("b1="+b1);
        System.out.println("b2="+b2);

        b1 >>=31;                              //b1 右移 31 位
        System.out.println("b1="+b1);
        b1 >>=1;                               //b1 再右移 1 位
        System.out.println("b1="+b1);

        b2 >>>=31;                             //b2 右移 31 位
        System.out.println("b2="+b2);
        b2 >>>=1;                              //b2 再右移 1 位
        System.out.println("b2="+b2);
    }
}
```

相应的执行结果如图 3-1 所示。

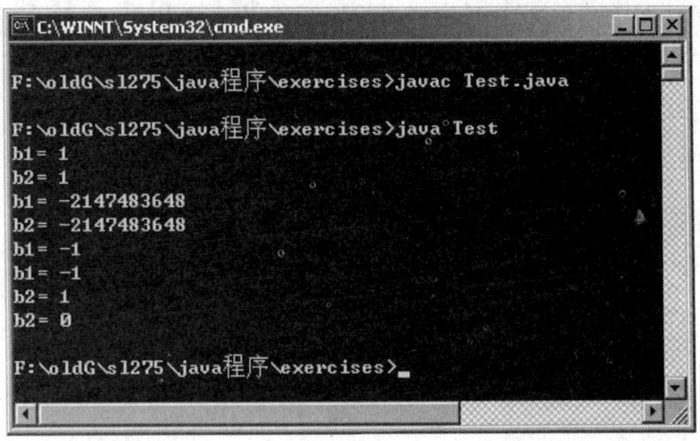

图 3-1 右移操作的执行结果

【拓展思考】

给出下面的说明,下列每个赋值语句会得到什么结果?

```
int iResult, num1=25, num2=40, num3=17, num4=5;
double fResult, val1=17.0, val2=12.78;
    a. iResult=num1/num4;
    b. fResult=num1/num4;
    c. iResult=num3/num4;
    d. fResult=num3/num4;
    e. fResult=val1/num4;
    f. fResult=val1/val2;
    g. iResult=num1/num2;
    h. fResult=(double) num1/num2;
    i. fResult=num1/(double) num2;
    j. fResult=(double) (num1/num2);
    k. iResult=(int) (val1/num4);
    l. fResult=(int) (val1/num4);
    m. fResult=(int) ((double) num1/num2);
    n. iResult=num3%num4;
    o. iResult=num2%num3;
    p. iResult=num3%num2;
    q. iResult=num2%num4;
```

3.4 设 n 为自然数,

$$n! = 1 \times 2 \times 3 \times \cdots \times n$$

称为 n 的阶乘,并且规定 $0!=1$。试编制程序计算 $2!,4!,6!,8!$ 和 $10!$,并将结果输出到屏幕上。

解:阶乘函数在数学上的定义为:

$$n! = \begin{cases} 1 & (n=0) \\ n(n-1)! & (n>0) \end{cases}$$

这是一个递归定义,因为阶乘本身又出现在阶乘的定义中。对于所有的递归定义,一定要有一个递归结束的出口,这既是定义的最基本情况,也是程序执行递归结束的地方。本定义中的第一行即是递归出口。

当一个函数使用递归定义的时候,往往直接使用递归方法实现它。

阶乘的递归实现如下所示:

```java
import java.util.*;

public class Factorial
{   public static void main(String[] args)
    {   Factorial ff=new Factorial();
        for (int i=0; i<5; i++)                    //共计算 5 个阶乘结果
        {   ff.setInitVal(2*(i+1));               //计算哪个值的阶乘
            ff.result=Factorial(ff.initVal);       //计算
            ff.print();                            //输出结果
        }
    }
    public static int Factorial(int n)
    {   if (n==0) return 1;                        //递归出口
        return n*Factorial(n-1);                   //递归计算
    }
    public void print()
    {   System.out.println(initVal+"!="+result);
    }
    public void setInitVal(int n)
    {   initVal=n;
    }
    private int result, initVal;
}
```

程序的执行结果如图 3-2 所示。

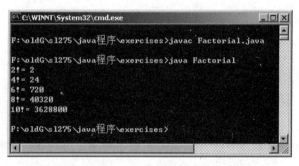

图 3-2　使用递归算法计算阶乘的执行结果

另一方面,根据阶乘的定义,$n!=n(n-1)(n-2)\cdots 3\times 2\times 1$,完全可以用循环来计算阶乘,而不必使用递归。因为递归毕竟多次调用同一个方法,函数调用所花的时间较长,递归调用的效率较低。

阶乘的非递归实现如下所示：

```java
import java.util.*;

class Factorial2
{   private int result, initVal;
    public static void main(String[] args)
    {   Factorial2 ff=new Factorial2();
        for (int i=0; i<5; i++)
        {   ff.setInitVal( 2 * (i+1));              //计算初值
            ff.result=1;                             //连乘运算的初值为1
            for (int j=2; j<=ff.initVal; j++)        //循环计算连乘结果
                ff.result *=j;
            ff.print();
        }
    }
    public void print()
    {   System.out.println(initVal+"!="+result);
    }
    public void setInitVal(int n)
    {   initVal=n;
    }
}
```

3.5 使用 java.lang.Math 类，生成 100 个 0～99 之间的随机整数，找出它们之中的最大者及最小者，并统计大于 50 的整数个数。

提示：

Math 类支持 random 方法：

```
public static synchronized double random()
```

该方法返回一个 0.0～1.0 之间的小数，如果要得到其他范围的数，需要进行相应的转换。例如想得到一个 0～99 之间的整数，可以使用下列语句：

```
int num=(int)(100 * Math.random());
```

解：提示中已经说明了，可以使用 Math.random() 方法得到随机数，但这个随机数是一个 0.0～1.0 之间的浮点数，首先需要将数的范围变化到 0～99 之间，然后再将得到的数转换为整数。

程序中，使用了两个变量 MAXof100、MINof100 分别记录这 100 个整数中的最大值和最小值。先生成前两个随机整数，较大者放入 MAXof100 中，较小者放入 MINof100 中。随后使用一个循环生成剩余的 98 个随机整数，然后分别与 MAXof100 和 MINof100 进行比较，新生成的数如果大于 MAXof100，则将 MAXof100 修改为新的数。同样如果新生成的数小于 MINof100，则让 MINof100 记下这个数。程序中使用 count 记录大于 50

的随机数的个数,初始时,它的值为 0。

程序代码实现如下:

```java
import java.util.*;

public class MathRandomTest
{   public static void main(String[] args)
    {   int count=0, MAXof100, MINof100;
        int num,i;

        MAXof100=(int)(100*Math.random());          //生成的第一个随机数
        MINof100=(int)(100*Math.random());          //生成的第二个随机数
        System.out.print(MAXof100+"   ");
        System.out.print(MINof100+"   ");
        if (MAXof100>50) count++;                   //记录下大于 50 的个数
        if (MINof100>50) count++;                   //记录下大于 50 的个数

        if (MINof100>MAXof100)                      //比较前两个随机数
        {   num=MINof100;
            MINof100=MAXof100;                      //较小者记入 MINof100
            MAXof100=num;                           //较大者记入 MAXof100
        }

        for (i=0; i<98; i++)                        //接下来生成其余的 98 个随机数
        {   num=(int)(100*Math.random());
            //控制每输出 10 个数即换行
            System.out.print(num+ ((i+2)%10==9 ?"\n" : "   "));
            if (num>MAXof100)
                MAXof100=num;                       //更大的数记入 MAXof100
            else if (num<MINof100)
                MINof100=num;                       //更小的数记入 MINof100
            if (num>50) count++;                    //记录下大于 50 的个数
        }
        System.out.println("The MAX of 100 random integers is:   "+MAXof100);
        System.out.println("The MIN of 100 random integers is:   "+MINof100);
        System.out.println("The number of random more than 50 is:   "+count);
    }
}
```

程序的执行结果如图 3-3 所示。

【拓展思考】

(1) 哪个包包含 Scanner 类? String 类又在哪个包内? Random 类呢? Math 类呢?

解: Scanner 类和 Random 类属于 java.util 包。String 和 Math 类属于 java.lang 包。

(2) 为什么不需要在程序中引入 Math 类?

解: Math 类属于 java.lang 包,这个包自动引入到任一个 Java 程序中,所以不需要

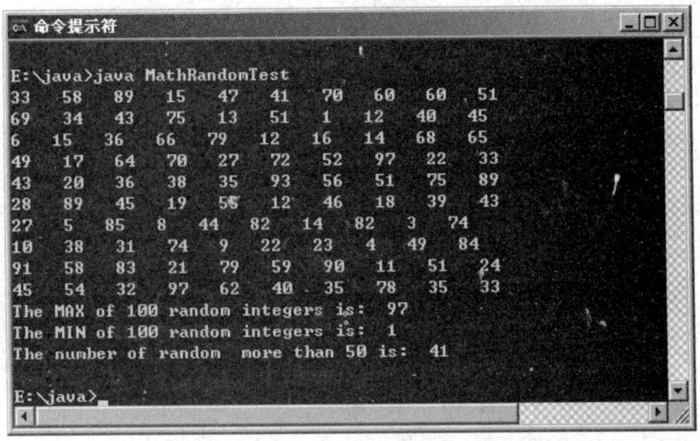

图 3-3 随机整数的执行结果

使用单独的 import 语句来说明。

(3) 给定一个 Random 对象 rand，调用 rand.nextInt()将返回什么？

解：调用 Random 对象的 nextInt()方法返回 int 值范围内的一个随机整数，包括正数和负数。

(4) 给定一个 Random 对象 rand，调用 rand.nextInt(20)将返回什么？

解：给 Random 对象的 nextInt()方法传递一个正整数参数 x，返回 $0 \sim x-1$ 范围内的一个随机数。所以调用 nextInt(20)将得到 0~19(含)之间的一个随机数。

(5) 写一个语句，打印 1.23 弧度的正弦值。

解：下列语句打印 1.23 弧度的正弦值：

```
System.out.println (Math.sin(1.23));
```

(6) 说明一个 double 类型变量 result，初始化为 $5^{2.5}$。

解：下列说明创建了变量 double，并初始化为 $5^{2.5}$。

```
double result=Math.pow(5, 2.5);
```

3.6 下列表达式中，找出每个操作符的计算顺序，在操作符下按次序标上相应的数字。

```
a+b+c-d
a+b/c-d
a+b/c*d
(a+b)+c-d
(a+b)+(c-d)%e
(a+b)+c-d%e
(a+b)%e%c-d
```

解：在 Java 中，在对一个表达式进行计算时，如果表达式中含有多种运算符，则要按运算符的优先顺序依次从高向低进行，同级运算符则从左向右进行。括号可以改变运算次序。运算符的优先次序参见 3.2 题答案。

各个表达式中运算符的优先次序如下：

```
a + b + c - d
  1   2   3

a + b / c - d
  2   1   3

a + b / c * d
  3   1   2

( a + b ) + c - d
    1     2   3

( a + b ) + ( c - d ) % e
    1     4     2     3

( a + b ) + c - d % e
    1     2   4   3

( a + b ) % e % c - d
    1     2   3   4
```

3.7 编写程序打印下面的图案。

```
*******
 *****
  ***
   *
  ***
 *****
*******
```

解：从图中可以看出，该图以中间行为基准，上下对称。首先看看要打印的总行数，如果每行都不同，则需要定制各行的打印内容。如果有重复或是对称内容的话，则可以简化语句的处理。观察本题中要输出的内容，一共要输出7行，以 initNum 变量来表示。第1行和第7行一样，第2行和第6行一样，第3行和第5行一样，实际上我们只需要定制4个不同的行即可。

从1开始计行号。各行的星号数与行号 i 的关系为：当 i<=(initNum+1)/2，相应行的星号数为：(initNum-2*(i-1))；当 i>(initNum+1)/2，相应行的星号数为：(2*i-initNum)，两个星号之间有两个空格。再看每行最左侧的空格数与行号的关系：当 i<=(initNum+1)/2，相应行的空格个数为：3(i-1)；当 i>(initNum+1)/2，相应行的空格数为：(21-3i)。另外定义两个函数 printaster() 和 printspace()，分别用来输出空格和星号。在程序实现时，每行先输出空格，后输出星号。

程序代码实现如下：

```
import java.util.*;

public class PrintAst
{   public static void main(String[] args)
```

```java
    {   PrintAst pa=new PrintAst();
        int initNum=7;                              //总共输出7行

        for(int i=1;i<=initNum; i++)
        {   if(i<=(initNum+1)/2)                    //前半部分
            {   for(int m=1; m<=3*(i-1); m++)
                {   pa.printSpace();                //输出空格
                }
                for(int k=1; k<=initNum-2*(i-1); k++)
                {   pa.printAstar();                //输出星号
                }
            }
            else                                    //后半部分
            {   for(int m=21-3*i; m>0; m--)
                {   pa.printSpace();
                }
                for(int k=1; k<=2*i-initNum; k++)
                {   pa.printAstar();
                }
            }
            System.out.print("\n");
        }
    }

    public void printAstar()
    {   System.out.print("* ");
    }
    public void printSpace()
    {   System.out.print(" ");
    }
}
```

3.8 编写程序打印下面的图案。

```
* * * * * * * * *
 * * * * * * * *
  * * * * * * *
   * * * * * *
    * * * * *
     * * * *
      * * *
       * *
        *
```

解:这个图案比 3.7 题的图案简单一些,因为它的变化是一致的,不需要分前后两部分。同样地,需要先判定输出的总行数,本题中是 10 行。接下来,确定每行输出的星号数及起始位置。从图中看出,起始位置都是从第一列开始。第一行输出 10 个星号,以后每行递减一个星号,直到最后一行仅输出一个星号。

程序代码实现如下:

```
public class PrintTriag
{   public static void main(String[] args)
    {   int initLine=10;                    //总共输出 10 行
        int initNum=10;                     //一行中最多的星号数
        PrintTriag pt=new PrintTriag();
        for (int i=0; i<initLine ; i++)
            for(int j=0; j<initNum-i; j++)
            {   pt.printAstar();
            }
            System.out.print("\n");
    }
    public void printAstar()
    {   System.out.print(" * ");
    }
}
```

3.9 编写程序打印乘法口诀表。

解:乘法口诀表是学习数学时入门级的知识,主要内容是 10 以内的两个数相乘的结果,按行输出。使用循环语句可以完成。输出的格式为:被乘数相同的结果输出在同一行中,乘数相同的结果输出在同一列中。

程序代码实现如下:

```
import java.util.*;

public class MultipleTable
{   public static void main(String[] args)
    {   MultipleTable mt=new MultipleTable();

        int initNum=9;                          //输出共 9 行
        int res=0;                              //计算乘法结果
        for(int i=1;i <=initNum; i++)           //行的控制
        {   for(int j=1;j <=i;j++)              //列的控制
            {   res=i*j;                        //乘积
                mt.printFormula(i, j, res);
            }
            System.out.print("\n");
        }
    }
```

```java
    public void printFormula(int i,int j ,int res)
    {   System.out.print( i+" * " +j +"="+res+"   " );
    }
}
```

程序输出结果如图 3-4 所示。

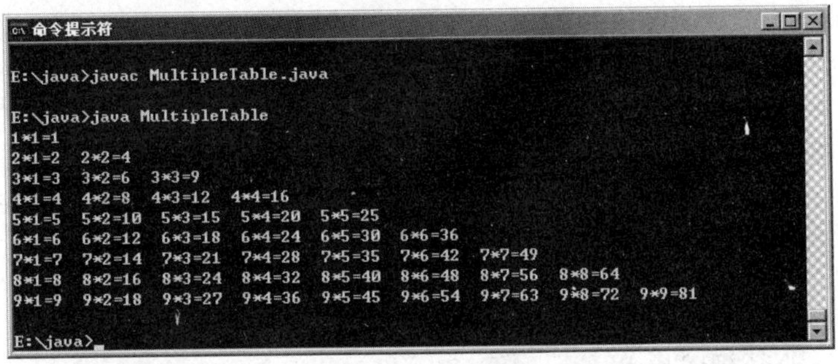

图 3-4 乘法口诀表

3.10 编写程序,要求判断从键盘输入的字符串是否为回文(回文是指自左向右读与自右向左读完全一样的字符串)。

解:如题中所说,回文即自左向右读与自右向左读完全一样的字符串。那么如何判断一个字符串是回文呢?有很多的方法,我们介绍其中比较简单的两种实现方法。

方法一:从字符串的两头相向比较,第一个字符与最后一个字符比较,第二个字符与倒数第二个字符比较,以此类推,直到字符串的中间位置为止。对中间位置字符的比较分两种情况,一是字符串中含有偶数个符号,此时刚好配对比较成功。另一种情况是字符串中含有奇数个符号,此时中间位置的符号不需要和任何符号进行比较。如果这些比较均相等,则字符串是回文。

方法二:设字符串为 w,将字符串全部反转变为 w1。例如字符串"abcdefg"反转后的结果是"gfedcba"。将 w 与 w1 进行比较,如果相等,则为回文。

从键盘输入字符串的方法,将在第 11 章详细介绍,本题中读者可以忽略此处。因为有标准输入/输出的操作,因此程序中对异常也进行了处理。异常处理是第 6 章中的内容。

方法一程序代码实现如下:

```java
import java.io.*;
public class HuiWen
{   boolean isHuiWen(char str[], int n)
    {   int net=0;                              //记录已经比较的符号数
        int i, j;                    //需要进行比较的左右标记,i 为左侧符号,j 为右侧符号
        for(i=0, j=n-1; i<n/2; i++, j--)
        {   if(str[i]==str[j]) net++;
        }
        if(net==(int)(n/2))                     //比较到了中间位置
```

```
            {   return true;
            }
            else
            {   return false;
            }
        }

        public static void main(String[] args)
        {   HuiWen hw1=new HuiWen();
            String pm="";
            Try                                           //异常处理
            {   InputStreamReader reader=new InputStreamReader(System.in);
                BufferedReader input=new BufferedReader(reader);
                System.out.print("give your test String :\n ");
                pm =input.readLine();
                System.out.println(pm);
            } catch(IOException e)                        //IOException 是个标准异常
            {   System.out.println("exception occur...");
            }

            boolean bw=hw1.isHuiWen(pm.toCharArray(), pm.length());
            if (bw==true)
            {   System.out.println("It is huiwen!");
            }
            else
            {   System.out.println("It is not huiwen");
            }
        }
    }
```

程序的执行结果如图 3-5 所示。

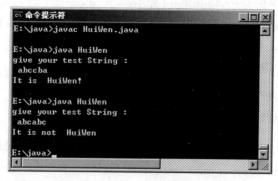

图 3-5　HuiWen 类的执行结果

第二种方法可以让读者熟悉字符串及字符数组的使用,其程序代码实现如下:

```java
import java.io.*;
public class HuiWen2
{
    String reverse(String w1)
    {
        String w2;
        char []str1=w1.toCharArray();              //字符串转成字符数组
        int len=w1.length();
        char []str2=new char[len];                 //分配一个新的字符数组
        for (int i=0; i<len; i++)
        {
            str2[i]=str1[len-1-i];                 //用来保存原字符串的反转
        }
        w2=new String(str2);                       //返回字符串型
        return w2;
    }

    public static void main(String[] args)
    {
        HuiWen2 hw1=new HuiWen2();
        String pm="";
        try
        {   InputStreamReader reader=new InputStreamReader(System.in);
            BufferedReader input=new BufferedReader(reader);
            System.out.print("give your test String :\n ");
            pm=input.readLine();
        } catch (IOException e)
        {   System.out.println("exception occur...");
        }

        String w2=hw1.reverse(pm);                 //字符串反转
        if (w2.compareTo(pm)==0)                   //字符串比较
        {   System.out.println("It is a HuiWen!");
        }
        else
        {   System.out.println("It is not a HuiWen!");
        }
    }
}
```

【拓展思考】

如何比较字符串相等？

解：使用 String 类的 equals 方法可以对字符串进行相等比较，方法返回一个布尔结果。String 类的 compareTo 方法也可用来比较字符串。它根据两字符串的大小关系，返回正数、0或是负数。

3.11 编写程序,判断用户输入的数是否为素数。

解：素数是只能被1和本身整除的整数。换句话说,除1和本身外,素数没有其他因子。这点可以作为判定素数的规则。对于一个整数 n,如果从 $2\sim n-1$ 之间的任何一个整数都不能整除 n,则 n 为素数。进一步的分析可知,仅需判定从 $2\sim\sqrt{n}$ 之间的任何一个整数都不能整除 n,则可判定 n 为素数。这样可以减少循环的执行次数。

据此编写判断程序如下:

```java
import java.io.*;

public class PrimeNumber
{   private int pm;
    public void setPm(int pm)
    {   this.pm=pm;
    }
    public boolean isPrime()                        //判断素数
    {   boolean bl=true;
        int i=2;
        for(i=2; i<=Math.sqrt(pm);)                 //循环判定有否因子
        {   if(pm%i==0)
               {   bl=false;
                   break;                           //如果存在因子,则跳出循环
               }
            else
               {   i++;                             //否则继续
               }
        }
        return bl;
    }

    public static void main(String args[])
    {   PrimeNumber prim=new PrimeNumber();
        int testNum=0;
        try{    InputStreamReader reader=new InputStreamReader(System.in);
                BufferedReader input=new BufferedReader(reader);
                System.out.println("give your test number : ");
                testNum=Integer.parseInt(input.readLine());
        } catch (IOException e)
          {   System.out.println("exception occur...");
          }
        prim.setPm(testNum);
        boolean bl=prim.isPrime();
        if(bl==true)
```

```
            {   System.out.print(testNum+"is a prime number\n");
            }
            else
            {   System.out.print(testNum+"is not a prime number\n");
            }
        }
    }
```

程序的执行结果如图 3-6 所示。

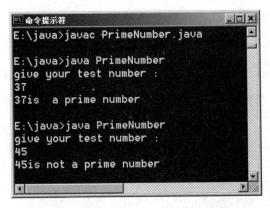

图 3-6 PrimeNumber 类的执行结果

【拓展思考】

(1) 程序中的控制流指什么?

解:对于程序的一次运行,程序的控制流决定着程序语句的执行。

(2) 条件语句与循环语句中的条件是基于什么类型的?

解:每种条件和循环语句都基于一个能得到真值或假值的布尔条件。

(3) 什么是嵌套的 if 语句? 什么是嵌套的循环语句?

解:当在 if 或是 else 子句中又出现 if 语句时,就是嵌套的 if 语句。嵌套 if 让程序做出一系列判别。同样地,嵌套的循环是在循环中又有循环。

(4) 在条件语句及循环语句中,如何使用块语句?

解:块语句将几个语句放到一起。当想基于布尔条件做多件事情时,可使用块语句来定义 if 语句或是循环语句的语句体。

(5) 在 switch 语句 case 分支的结尾处如果不使用 break 语句会发生什么?

解:如果多分支的 case 语句的最后没有 break 语句,则程序流将继续执行到下一个 case 的语句中。通常使用 break 语句,为的是跳到 switch 语句的结尾。

(6) 什么是相等运算符? 还有哪些关系运算符?

解:相等运算符是等于(==)和不等于(!=)。关系运算符还有小于(<)、小于等于(<=)、大于(>)和大于等于(>=)。

(7) 当对浮点数进行相等比较时,为什么要非常小心?

解:因为在计算机内部它们用二进制来存储,只有各个位都相同时,两个浮点值进行精确比较时才为真。所以最好使用合理的公差值来考虑两值之间的误差。

（8）列出 while 语句与 do 语句的相同与不同之处。

解：while 循环先计算条件。如果为真，则执行循环体。do 循环先执行循环体，再计算条件。所以 while 循环的循环体执行 0 次或多次，而 do 循环的循环体执行 1 次或多次。

（9）何时可用 for 循环代替 while 循环？

解：当知道或能够计算循环体的迭代次数时，常使用 for 循环。while 循环处理更一般的情形。

3.12 编写程序，将从键盘输入的华氏温度转换为摄氏温度。

解：华氏温度和摄氏温度之间的转换关系为：

Celsius=(Fahrenheit-32)/9*5。

同样地，因为要处理键盘输入，所以添加了异常处理。转换程序代码实现如下：

```
import java.io.*;
public class TempConverter
{   double celsius(double y)                    //转为摄氏温度
    {   return ((y-32)/9*5);
    }

    public static void main(String[] args)
    {   TempConverter tc=new TempConverter();
        double tmp=0;
        try
        {   //定义输入源
            InputStreamReader reader=new InputStreamReader(System.in);
            BufferedReader input=new BufferedReader(reader);
            System.out.print("give your fahrenheit temperature :\n ");
            //从键盘输入一行
            tmp=Double.valueOf(input.readLine());    //将输入转为双精度数
        }
        catch(NumberFormatException fe)
        {   System.out.println("format error...");
        }
        catch (IOException e)
        {   System.out.println("IoException occur...");
        }
        System.out.println("the celsius of temperature is"+tc.celsius(tmp));
    }
}
```

程序的执行结果如图 3-7 所示。

3.13 编写程序，读入一个三角形的三条边长，计算这个三角形的面积，并输出结果。

图 3-7　Temconverter 类的执行结果

提示：设三角形的三条边长分别是 a、b、c，则计算其面积的公式为：
$$s = (a+b+c)/2$$
$$面积 = \sqrt{s(s-a)(s-b)(s-c)}$$

解：根据平面几何的理论，当三个数能够满足任两数之和大于第三个数及任两数之差小于第三个数时，这三个数才能构成三角形的三条边。由三条边的逻辑关系，当上述条件满足一个时，另外一个也必然满足。因此只需要判定任两边之和大于第三边即可。根据题意，程序中定义了 Trigsquare 类，该类包括三角形的三条边这三个属性，以及判断是否能构成三角形的 isTriangle() 方法和求解三角形面积的 getArea() 方法。

程序代码实现如下：

```java
import java.io.*;
public class Trigsquare
{   double x;                                         //x、y、z 分别为 3 条边
    double y;
    double z;
    Trigsquare (double x, double y, double z)
    {   this.x=x;
        this.y=y;
        this.z=z;
    }

    boolean isTriangle()
    {   boolean bl=false;
        if(this.x>0 && this.y>0 && this.z>0)          //必须是 3 个正数,边长不能是负值
        {   if ((this.x+this.y)>this.z &&             //任两边之和大于第三边
            (this.y+this.z)>this.x && (this.x+this.z)>this.y)
                bl=true;                              //能够构成三角形的三条边
            else
                bl=false;                             //不能构成三角形的三条边
        }
        return bl;
    }
    double getArea()
```

```
{   double s=(this.x+this.y+this.z)/2.0;
    return (Math.sqrt(s * (s-this.x) * (s-this.y) * (s-this.z)));
}

public static void main(String[] args)
{   double s[]=new double[3];                    //使用一个数组来存放 3 个数
    try
    {   InputStreamReader reader=new InputStreamReader(System.in);
        BufferedReader input=new BufferedReader(reader);
        System.out.print("give your three number :\n ");
        for (int i=0; i<3; i++)                  //输入 3 个数,每行输入一个
        {   s[i]=Double.valueOf(input.readLine());
        }
    }catch(NumberFormatException fe)
    {   System.out.println("format error...");
    }catch(IOException e)
    {   System.out.println("IOException occur...");
    }
    Trigsquare ts=new Trigsquare(s[0], s[1], s[2]);
    if(ts.isTriangle()==true)
    {   System.out.println("The square of Triangle is "+ts.getArea());
    }
    else
    {   System.out.println("the numbers input cannot be a trangle!");
    }
}
}
```

程序的执行结果如图 3-8 所示。

图 3-8 Trigsquare 类的执行结果

3.14 编写一个日期计算程序,完成以下功能:

(1) 从键盘输入一个月份,屏幕上输出本年这个月的月历,每星期一行,从星期日开始,到星期六结束。

(2) 从键盘输入一个日期,屏幕上回答是星期几,也以当年为例。

(3) 从键盘输入两个日期,计算这两个日期之间含有多少天。

解:程序中用户可以任意选择实现上述三种功能。

Calendar 类是系统提供的关于日期的抽象类,它的具体子类 GregorianCalendar 类实现了一些基本的方法。本例从 GregorianCalendar 派生类 MyCalender,创建它的实例以得到某个月的月历。计算星期几则采用 Calendar 类中定义的 DAY_OF_WEEK 常量来得到,由于 Java 标准库中所有时间类都是以毫秒数为基础的,因而对两个日期的天数差值计算可以按如下方法进行:

```
diff=(date1.getTime()-date2.getTime())/(24*3600*1000)
```

程序代码实现如下:

```java
import java.util.*;
import java.io.*;
import java.util.Date.*;
import java.text.*;
import java.text.DateFormat.*;
import java.text.SimpleDateFormat.*;

public class MyCalender extends GregorianCalendar
{    void showCalender(int month)
    {    int[]days={31,28,31,30,31,30,31,31,30,31,30,31};//非闰年的每个月的天数
        Calendar c=Calendar.getInstance();
        int year=c.get(Calendar.YEAR);
        int date=c.get(Calendar.DATE);
        c=new GregorianCalendar(year, month-1, date);

        if((year%4==0 && year%100!=0||year%400==0) && month==2) days[1]++;
        //闰年的2月份天数
        System.out.println("------Calendar:"+year+"-"+month+"------\n");
        System.out.println("SUN MON TUE WED THU FRI SAT");
        c.set(DATE, 1);
        int first=c.get(Calendar.DAY_OF_WEEK);          //1日在星期几
        int i;
        for(i=1; i<first; i++) System.out.print("    ");  //1日之前的空白
        for(i=1; i<=days[month-1]; i++)
        {    if(i<10) System.out.print(" ");
            System.out.print(" "+i);
            System.out.print("  ");
            if(first++%7==0) System.out.println(" ");     //每星期占一行
        }
        System.out.println("\n");
    }
    String getDay(int month, int date)
    {    Calendar c=Calendar.getInstance();
```

```java
        int year=c.get(Calendar.YEAR);
        c=new GregorianCalendar(year, month-1, date);
        int day=c.get(Calendar.DAY_OF_WEEK);          //返回该日期是星期几
        String reday="";
        switch (day)
        {   case 1:
                reday="Sunday"; break;
            case 2:
                reday="Monday";break;
            case 3:
                reday="Tuesday";break;
            case 4:
                reday="Wednesday";break;
            case 5:
                reday="Thursday"; break;
            case 6 :
                reday="Friday" ;break;
            default:
                reday="Saturday";
        }
        return reday;
    }

    int  diffDate(java.util.Date date1, java.util.Date date2)
                                            //计算两个日期之间相差的天数
    {   return Math.abs((int)((date1.getTime()-date2.getTime())/(24*3600*1000)));
    }
    int inputNum()                          //输入选项号
    {   int pm=0;
        try
        {   InputStreamReader reader=new InputStreamReader(System.in);
            BufferedReader input=new BufferedReader(reader);
            pm=Integer.parseInt(input.readLine());
        }catch(NumberFormatException ne)
        {   System.out.println("invalid data format...");
        }catch (IOException e)
        {   System.out.println("exception occur...");
        }
        return pm;
    }
    String inputDate()                      //输入日期
    {   String pm="";
        try
        {   InputStreamReader reader=new InputStreamReader(System.in);
            BufferedReader input=new BufferedReader(reader);
            pm =input.readLine();
```

```java
        }catch (IOException e)
        {   System.out.println("exception occur...");
        }
        return pm;
    }

    public static void main(String[]args)
    {   MyCalender mc=new MyCalender();
        int count=1;
        int mon,date2;
        boolean flag=true;
        while(count!=0)
        {   System.out.print("which of following do you want to do:(1,2,3,0)\n ");
            System.out.print("1.show calendar.\n ");
            System.out.print("2.get the day of the date.\n ");
            System.out.print("3.count the days between two dates.\n ");
            System.out.print("0.exit.\n ");
            count=mc.inputNum();                    //选项号,有4种
            switch (count)
            {   case 1:                              //显示日期
                    System.out.print("input the month(1-12) : ");
                    mon=mc.inputNum();
                    if(mon <=12 && mon>0)           //月份要合理
                    {   System.out.print("\n");
                        mc.showCalender(mon);
                    }
                    break;
                case 2:                              //判定输入的日期是星期几
                    System.out.print("input the month: ");
                    mon=mc.inputNum();
                    System.out.print("input the date: ");
                    date2=mc.inputNum();
                    System.out.println("the day of the input date is "
                        +mc.getDay(mon,date2));
                    System.out.print("\n");
                    break;
                case 3:                              //计算两个日期之间的天数
                    java.util.Date mydate1=new java.util.Date();
                    java.util.Date mydate2=new java.util.Date();
                    java.text.DateFormat myFormatter=new
                        java.text.SimpleDateFormat("yyyy-MM-dd");
                    try
                    {   System.out.print("please input  date
                            one(format:year-mon-day) : ");
                        String d1=mc.inputDate();
                        System.out.print("please input  date
```

```
                two(format:year-mon-day): ");
                String d2=mc.inputDate();
                mydate1=myFormatter.parse(d1);
                mydate2=myFormatter.parse(d2);
            }catch(ParseException pe)
            {   System.out.println(pe);
            }
            System.out.println("the days between above are:"
                +mc.diffDate(mydate1,mydate2));
            break;
        case 0:                                    //退出
            System.exit(0);
        default:
            break;
        }//end of switch
    }//end of while
  }
}
```

程序的执行结果如图 3-9、图 3-10、图 3-11 所示。

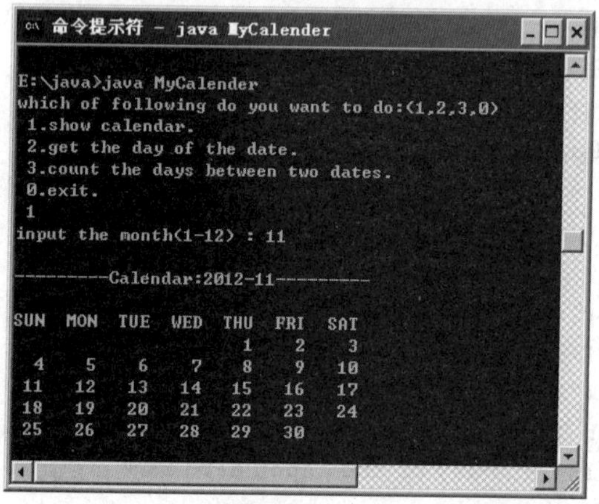

图 3-9　显示月历

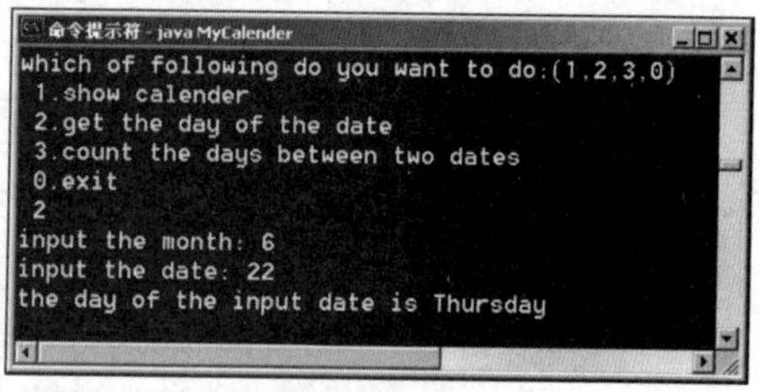

图 3-10　输出星期几

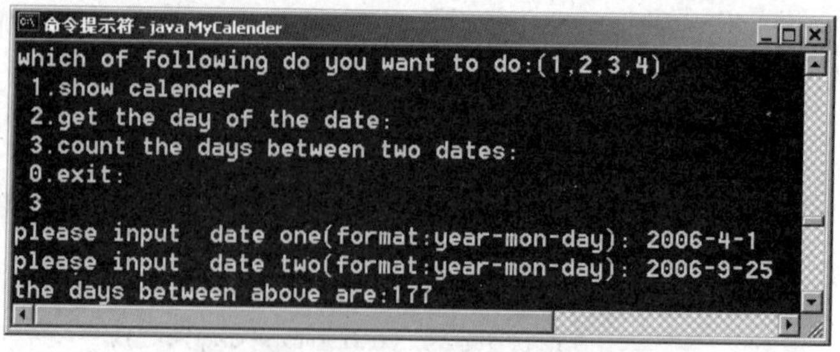

图 3-11　计算两日期间隔的天数

3.15 设有各不同面值人民币若干，编写一个计算程序，对任意输入的一个金额，给出能组合出这个值的最佳可能，要求使用的币值个数最少。例如，给出 1.46 元，将得到下列结果：

1.46 元＝

1 元　1 个

2 角　2 个

5 分　1 个

1 分　1 个

解：在实际生活中，由于人民币面值本身的特点，可以采用贪心算法。只要按照货币单位从大到小的次序来组合所给出的人民币，其结果货币个数就能达到最少。这里假设我们现有的人民币有如下面值：1 元、5 角、2 角、1 角、5 分、2 分、1 分。

程序代码实现如下：

```java
import java.io.*;

public class Moneycount
{   public static void main(String[] args)
    {   int[] val={100, 50, 20, 10, 5, 2, 1};   //分别对应的钱币的值,以分为单位
        int count;                              //钱币个数
        int [] num=new int[7];                  //各种钱币的数量
        int asval=0;
        System.out.println("请输入人民币币值(以分计):");
        try
        {   InputStreamReader isr=new InputStreamReader(System.in);
            BufferedReader br=new BufferedReader(isr);
            asval=Integer.parseInt(br.readLine());
        }catch( NumberFormatException ne)
        {   System.out.println("输入的不是数值!");
        }
        catch(IOException e)
        {   e.printStackTrace();
        }
        if(asval!=0)
        {   int reval=asval;
            for (int i=0; i<7; i++)
            {   num[i]=asval/val[i];
                asval=asval%val[i];
            }

            System.out.println("输入的人民币币值为"+reval);
            for(int i=0; i<7; i++)
            {   if(num[i]!=0)
```

```
                System.out.println(val[i]+"分的个数:"+num[i]);
            }
        }
    }
}
```

程序的执行结果如图 3-12 所示。

图 3-12　Moneycount 类的执行结果

第4章 数组和字符串

4.1 在 Java 中是如何完成数组边界检查的?

解:在 Java 中,数组下标从 0 开始,数组中的元素个数记录在变量 length 中,这是数组对象中唯一的数据成员变量。使用 new 创建数组时系统自动给 length 赋值。数组一旦创建完毕,其大小就固定下来。程序运行时可以使用 length 进行数组边界检查。如果发生越界访问,则抛出一个异常。

4.2 请简述数组创建的过程。如何创建一个对象数组?

解:数组的创建分两个步骤,一是定义,二是初始化。定义一个数组只是对数组的说明,此时系统还没有为数组分配任何内存空间,因此我们还不能访问它的任何元素。必须经过数组初始化后,才能应用数组的元素。定义加初始化这个过程就是数组的创建过程。

从数组的创建过程中可以看出,初始化是很关键的一步。数组的初始化分为静态初始化和动态初始化两种。所谓静态初始化就是在定义数组的同时对数组元素进行初始化,静态初始化可用于任何元素类型,初值块中每个位置的每个元素对应一个引用。

数组静态初始化的创建格式如下:

```
type arrayName[ ]={val1, val2, val3, val4…};    //val1, val2…为初始化的值
```

将需要赋给数组元素的值列在大括号中,并以逗号分隔。

与之相对应的,动态初始化是使用运算符 new 为数组分配空间,这和所有对象是一样的。数组说明的方括号中的数字表示数组元素个数。

对于简单类型的数组,其创建格式有以下两种:

```
type arrayName[ ]=new type[arraySize];
type[ ] arrayName=new type[arraySize];
```

如果前面已经对数组进行了说明,则此处的 type 可以不写。如下所示:

```
s=new char[20];
```

这一行创建了有 20 个字符的数组。

对于复合类型的数组,使用运算符 new 只是为数组本身分配空间,并没有对数组的元素进行初始化。所以对于复合类型的数组,需要经过两步进行空间分配。

第一步先创建数组本身:

```
type arrayName[ ]=new type[arraySize];
```

第二步分别创建各个数组元素:

```
arrayName[0]=new type(paramList);
```

```
...
arrayName[arraySize-1]=new type(paramList);
```

4.3 数组的内存分配是如何完成的？

解：系统在初始化时为数组分配内存，Java 中没有静态的数组定义，数组的内存都是通过 new 动态分配的。参考习题 4.2。

4.4 下列哪个语句段可以生成含 5 个空字符串的数组？

(1)

```
String a []=new String [5];
for (int i=0; i<5; a[i++]="");
```

(2)

```
String a []={"", "", "", "", ""};
```

(3)

```
String a [5];
```

(4)

```
String [5] a;
```

(5)

```
String [] a=new String [5];
for (int i=0; i<5; a[i++]=null);
```

解：将上述说明及赋值语句放到 main 函数中进行测试，为了区别，我们使用 a1 到 a5 分别命名各个数组。程序的最后将数组的各个元素输出，每个数组占用一行，以@@作为数组元素之间的分隔符。

程序代码实现如下：

```
import java.util.*;

public class ArrayValue
{   public static void main(String[] args)
    {   String a1[]=new String [5];
        for (int i=0; i<5; a1[i++] ="");
        String a2[]={"","","","",""};
        //String a3[5];
        //String [5] a4;
        String [] a5=new String [5];
        for (int i=0; i<5; a5[i++]=null);

        for (int i=0; i<5; i++)
            System.out.print("a1["+i+"]="+a1[i]+" @@");
```

```
        System.out.println();

        for (int i=0; i<5; i++)
            System.out.print("a2["+i+"]="+a2[i]+" @@");
        System.out.println();

        for (int i=0; i<5; i++)
            System.out.print("a5["+i+"]="+a5[i]+" @@");
        System.out.println();
    }
}
```

a1 和 a2 都是含 5 个元素的数组,数组的每个元素都是空字符串。Java 中没有静态数组,所以第 3 种和第 4 种的说明是错误的。编译时,系统会显示如图 4-1 所示的错误信息。另外,给 a5 的各元素赋值 null,并不表示是空字符串,输出 a5 的结果时可以看出这一点。

将 a3 及 a4 的说明去掉,再次编译并执行程序,得到如图 4-2 所示的结果。

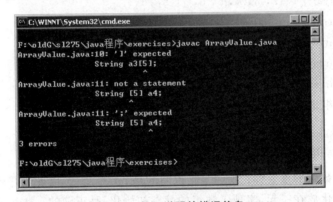

图 4-1　数组说明的错误信息

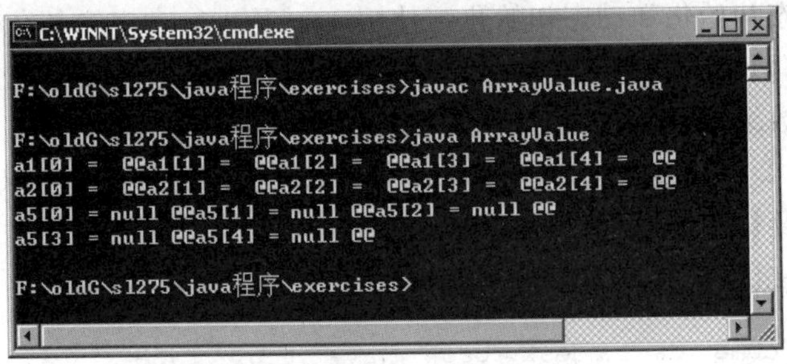

图 4-2　数组说明的执行结果

4.5 下面的数组定义哪些是正确的？

(1) int a[ ][ ]=new int [10, 10];

(2) int a[10][10]=new int [ ][ ];

(3) int a[ ][ ]=new int [10][10];
(4) int [ ]a[ ]=new int [10][10];
(5) int [ ][ ]a=new int [10][10];

解：从所给的5种说明中可以看出，各语句都想说明一个int型的二维数组。二维数组的一般说明格式有以下三种：

type arrayname[ ][ ]=new type [length1][length2];
type [ ]arrayname [ ]=new type [length1][length2];
type [ ][ ]arrayname=new type [length1][length2];

按照这个格式，第(1)种情况，等号右侧的维数说明应该改为[10][10]。第(2)种情况，等号左侧不允许出现维数说明，也就是不能有静态数组说明。其余三行的数组说明都是正确的，它们都能说明一个二维的整数数组，每一维含10个元素，每个数组中共有100个元素。

4.6 下面的数组定义哪些是正确的？能够实例化一个对象数组的是哪些？

(1) int array1={2,3,4,5,6,7};
(2) int array2[ ]= {2,3,4,5,6,7,8};
(3) int[ ] array3=int[30];
(4) int [ ] array4=new int [30];
(5) int [ ] array5=new {2,3,4,5,6,7,8};
(6) int [ ] array6=new int[ ];

解：一维数组定义的格式主要有以下两种：

type arrayName[ ];
type[ ] arrayName;

第1行没有数组的重要标记"[ ]"，它说明的是int型变量，因此不是数组。第2行将数组说明及初始化放在一起，实现了数组的静态初始化，是正确的。数组的动态初始化是使用运算符new为数组分配空间，所以第3行不正确。对应地，第4行是正确的；第5行混淆了数组的静态初始化和动态初始化。使用new是进行动态初始化，使用大括号是进行静态初始化，这两者不能混用。初始化时使用new只是为数组分配空间，类型后方括号中的数字表示数组元素个数，并且不能省略，所以第6行也是错误的。在上面的数组定义中，正确的是第2行和第4行。能够实例化一个对象数组的是第2行和第4行，因为它们都完成了定义及初始化过程。

4.7 选择一组等长的英文单词。例如，一组4个字母组成的单词：

work back come deal desk book java tool face

一组5个字母组成的单词：

watch match noise risky stock

试定义一个字符串数组，数组中每个元素存储一个英文单词，元素个数根据选择的英语单词数量而定。再按照电话机表盘定义数字与字母的对应关系，如数字2对应a或b

或 c,数字 5 对应 j 或 k 或 l。现编制一个程序,要求将用户输入的数字串转换成相应的字符串(注意一个数字串对应多个字符串),将这些字符串与数组中存储的英文单词逐个比较。如果某一字符串与英文单词匹配成功,则在屏幕上输出数字串及对应的单词。如果都不匹配,则在屏幕上输出一条信息"没有匹配的单词"。

解:程序中只建立了一个小的数组,存储的单词数量有限。读者可以自己扩充这部分,建立一个较大的词典,以便能匹配更多的词。

程序代码实现如下:

```java
import java.io.*;
import java.util.*;
import java.lang.*;
import java.lang.String;

class word                                          //word 对象
{   char[] invmap={'2','2','2','3','3','3','4','4','4','5','5','5',
        '6','6','6','7','7','7','7','8','8','8','9','9','9','9'};
    String wd;                                      //保存字母
    String nwd;                                     //保存字母数

    word(String s)                                  //构造方法
    {   wd=s;
        getNum();
    }

    void getNum()                                   //将字母转为数字
    {   char[] temp=wd.toCharArray();
        for(int i=0; i<wd.length(); i++)
        {   temp[i]=invmap[ (int) wd.charAt(i)-97];
            String ts=new String(temp);
            nwd=ts;
        }
    }
}//class word

class wordLib                                       //小字典库
{
    LinkedList WordList=new LinkedList();
    wordLib()
    {}
    void ConstructLib(String[] WordArray)
    //从字符串数组构造字典库
    {   for(int i=0; i<WordArray.length; i++)
        {   word temp=new word(WordArray[i]);
```

```java
            ListIterator iter=WordList.listIterator();
            if(WordList.isEmpty())
                WordList.addFirst(temp);
            else
            {   int addf=1;
                while(iter.hasNext())
                {   word t=(word)iter.next();
                    if(t.nwd.compareTo(temp.nwd) >=0)
                    {   if(iter.hasPrevious())
                        {   iter.previous();
                            iter.add(temp);
                            addf=0;
                            break;
                        }
                        else
                        {   WordList.addFirst(temp);
                            addf=0;
                            break;
                        }
                    }
                }                                      //while end
                if( addf==1)
                    WordList.addLast(temp);
            }
        }
    }                                              //构建字典库结束
}

class ResultList                                   //结果列表
{   LinkedList result=new LinkedList();

    void add(word wd)
    {   result.add(wd);
    }
}

public class findword                              //main class
{   static String[] FiveWord={"watch", "match", "noise", "risky", "stock"};
    static String[] FourWord={"work", "back", "come", "deal", "desk",
            "book", "java", "tool", "face"};

    public static void main(String[] args)
    {   boolean flag=true, errflag=false;
        String s="";                               //读输入字符串
```

```java
wordLib wl=new wordLib();
wl.ConstructLib(FiveWord);
wl.ConstructLib(FourWord);                    //构造字典库
ListIterator iter=wl.WordList.listIterator();
try
{   InputStreamReader isr=new InputStreamReader(System.in);
    BufferedReader br=new BufferedReader(isr);

    while(flag)                               //while 循环用于读入用户输入
    {   System.out.println(
            "please input your number, input 'exit' to quit");
        s=br.readLine();
        if(s.compareToIgnoreCase("exit")==0)
        {   flag=false;
            System.exit(0);
        }
        for(int i=0; i<s.length(); i++)
        {   if(s.charAt(i)>'9'||s.charAt(i)<'2')
            {   errflag=true;
                System.out.println(
                    "the input must be digital between 2 to 9");
                break;
            }
        }                                     //end of read input
        ResultList re=new ResultList();
        //查找匹配的字符串
        ListIterator iterator=wl.WordList.listIterator();
        while(iterator.hasNext())
        {   word tm=(word)iterator.next();
            if(tm.nwd.compareTo(s)==0)
            {   re.add(tm);
            }
            if(tm.nwd.compareTo(s)>0)
            break;
        }

        ListIterator reIter=re.result.listIterator();
        //输出结果

        if(re.result.isEmpty() && (!errflag))
        {   System.out.println("there is no matching word.");
        }
        whild (!re.result.isEmpty() && reIter.hasNext())
        {   word tm=(word)reIter.next();
```

```
                System.out.println("result is: "+tm.nwd+"    "+tm.wd);
            }
            errflag=false;
        }
    }catch(IOException e)
    {    e.printStackTrace();
    }
    }    //end of main fuction
}    //end of findword class
```

程序的执行结果如图4-3所示。

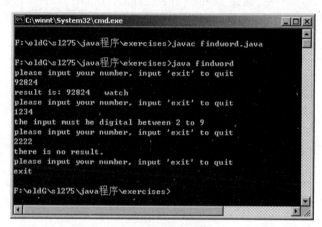

图4-3 单词匹配示例

4.8 定义一个一维的整数数组,其中存储随机生成的100个整数。利用你所熟悉的一种排序方法对它们进行升序排序,输出排序后的结果。

解:分别选择两种排序方法,一种是简单选择排序算法,另一种是起泡排序算法,分别实现该排序过程。

简单选择排序每次从剩余数据中选择最小值放到最前面,然后缩小范围,再从剩余数据中选择最小值放到本范围内的最前面。继续这个过程,范围越来越小,直到最后只剩一个数据时为止。

起泡排序是从数组的最后端开始向前扫描。每趟排序时,如果相邻两个数据呈逆序,则调整之。这样,一趟扫描后,最小值被调整到最前端。然后再从数组的最后端开始,继续上述过程,每一趟扫描都将本次范围内的最小者放到本范围的最前位置。也就是依次将数组中的最小值、次小值……调整到数组的最前端、第二个位置……。

(1)采用简单选择排序算法实现

```
public class ArraySort
{   public void sortArr(int [] arr,int len)
    {   //简单选择排序
        int k=0, tmp=0;
        for (int i=0;i<len-1 ;i++)
```

```
        {   k=i;
            for (int j=i+1;j<len;j++)
            {   if(arr[j]<arr[k])
                {   k=j;                        //用 k 选取每一趟中的最小值
                }
            }
            tmp=arr[i];                         //arr[j]与 arr[i]交换
            arr[i]=arr[k];
            arr[k]=tmp;
        }
    }
    public void print(int [] arr,int len)
    {   for(int i=1;i<=len;i++)
        {   System.out.print(arr[i-1]+"\t");
        }
    }

    public static void main(String[] args)
    {   ArraySort as=new ArraySort();
        int arr[]=new int[100];
        System.out.print("随机生成的整数数组的值如下:\n");
        for (int i=0;i<100 ;i++)
        {   arr[i]=(int)(100*Math.random());
            System.out.print(arr[i]+"\t");
        }

        //排序
        as.sortArr(arr,100);
        System.out.print("排序后的整数数组的值如下:\n");
        as.print(arr,100);
    }
}
```

(2) 采用起泡排序算法实现

```
public class  ArraySort2
{   public void sortArr2(int [] arr,int len)
    {   //起泡排序
        int tmp=0;
        for (int i=0;i<len;i++)
        {   for (int j=len-1;j>=i+1;j--)
            {   if(arr[j]<arr[j-1])
                {   tmp=arr[j];                 //arr[j]与 arr[i]交换
                    arr[j]=arr[j-1];
                    arr[j-1]=tmp;
```

```
            }
          }
        }
    }
    public void print(int [] arr,int len)
    {   for(int i=1;i<=len;i++)
        {   System.out.print(arr[i-1]+"\t");
        }
    }

    public static void main(String[] args)
    {   ArraySort2 as=new ArraySort2();
        int arr[]=new int[100];
        System.out.print("随机生成的整数数组的值如下:\n");
        for (int i=0;i<100 ;i++)
        {   arr[i]=(int) (100*Math.random());
            System.out.print(arr[i]+"\t");
        }

        //排序
        as.sortArr2(arr,100);
        System.out.print("排序后的整数数组的值如下:\n");
        as.print(arr,100);
    }
}
```

程序的执行结果如图 4-4 所示。

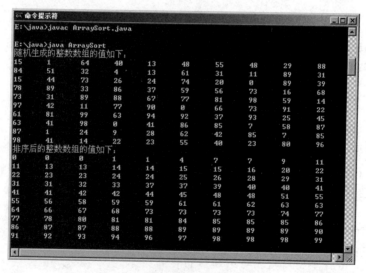

图 4-4　随机数组的排序执行结果

4.9 定义一个一维数组,其中存储随机生成的 1000 个 1~100 以内的整数,统计每个整数出现的次数。

解：首先定义一个含 1000 个元素的一维数组 arr，用来保存生成的随机数。生成随机数的方法是 Math.random()，数的范围是 0~1 之间的浮点数。如果要生成 100 以内的随机整数的话，计算公式为：(int)(100 * Math.random()+1)。

之后定义一个含 100 个元素的一维数组 arrCount，存储每个整数出现的次数。arrCount 数组中每个元素的初始值为 0，程序代码实现如下：

```java
public class ArrayCount
{   public static void main(String[] args)
    {   int arr[]=new int[1000];                    //存储随机数的数组
        for (int i=0;i<1000 ;i++)
        {   arr[i]=(int)(100 * Math.random()+1);    //随机生成 1 至 100 的整数
        }

        int arrCount[]=new int[100];                //用来记数的数组
        for (int i=0;i<100 ;i++)                    //数组初始化
        {   arrCount[i]=0;
        }

        for (int i=0;i<1000 ;i++)
        {   arrCount[arr[i]-1]++;                   //对应的个数加 1
        }
        System.out.print("整数数组的各值的个数如下:\n");
        for(int i=1;i<=arrCount.length;i++)
        {   if(arrCount[i-1]!=0)
            System.out.print((i)+"的个数为:"+arrCount[i-1]+"\t");
        }
        System.out.print("\n");
    }
}
```

程序执行结果如图 4-5 所示。

4.10 定义一个 Student 数组，其中保存学生的基本信息，包括姓名、学号、性别，还分别保存 3 门课程的成绩及 3 门课程对应的学分，试编程计算这 3 门课程的学分绩，并按学分绩的降序进行排序，输出排序后的结果。

解：根据题意，共有两种对象，一是课程，二是学生。因此分别定义两个类。对于课程对象，定义了 sclass 类，用来存储课程的基本信息，包括课程编号 classid 及其对应的学分 credit；对于学生对象，定义了 Student 类，存储学生的基本信息，包括学号 sid、姓名 sname、性别 sex，另外还有 3 门课程的成绩及学分绩 avgscore，3 门课程的成绩存储在数组 score 中。为了测试程序，定义了主类 Stuscore，这个类中包含学分绩的计算方法 accuscore()和按照学分绩对学生进行降序排序的 sort()方法。多名学生的信息保存在一个数组中。

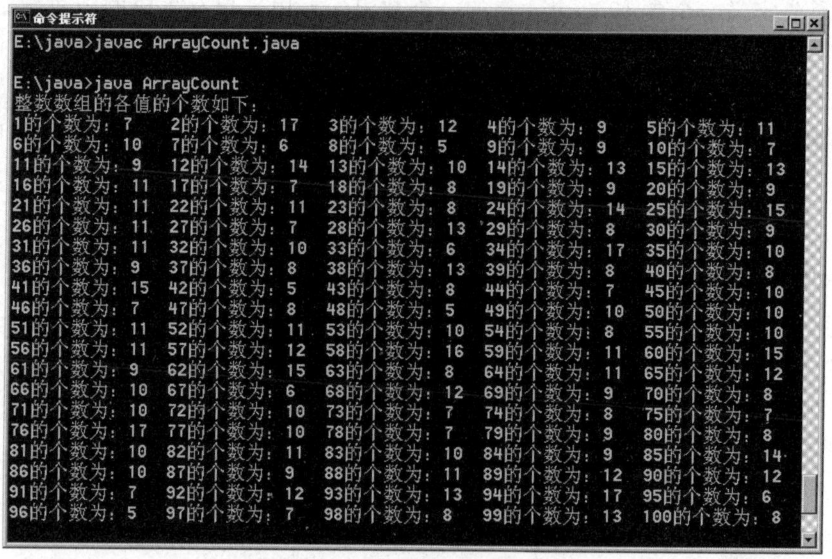

图 4-5 统计随机数各值的个数

程序代码实现如下：

//定义一个 Student 数组，其中保存学生的基本信息，包括学号、姓名、性别，
//还分别保存 3 门课程的成绩及 3 门课程对应的学分。
//程序可以计算这 3 门课程的学分绩，并按学分绩的降序进行排序，输出排序后的结果。

```
class sclass
{   String classid;                                    //课程编号
    int credit ;                                       //学分
    sclass(String classid, int credit)
    {   this.classid=classid;
        this.credit=credit;
    }
};

class Student
{   String sid;                                        //学号
    String sname;                                      //姓名
    char sex;                                          //性别 m-男生 ,f-女生
    float avgscore;                                    //学分绩
    float []score=new float[3];                        //成绩

    Student (String id, String name, char sex, float s1, float s2, float s3)
    {   this.sid=id;
        this.sname=name;
        this.sex=sex;
```

```java
        this.score[0]=s1;
        this.score[1]=s2;
        this.score[2]=s3;
    }
    Student(){};
    Student (String id, String name, char sex)
    {   this.sid=id;
        this.sname=name;
        this.sex=sex;
    }
    void print()
    {   System.out.println("id:"+this.sid+"\t\t name:"+this.sname+"\t sex:"
        +this.sex);
        System.out.println("score1="+this.score[0]+"\t score2="
            +this.score[1]+"\t sccore3="+this.score[2]);
        System.out.println("average score:"+this.avgscore);
        System.out.print("\n");
    }
    void print2()
    {   System.out.println("student id:"+this.sid+"\t average score:"+this.avgscore);
    }
};

public class Stuscore
{   Student []st;                                    //学生数组的定义
    sclass   []sc=new sclass[3];                     //3门课程
    void initsc()                                    //课程数组初始化
    {   sc[0]=new sclass("C001", 2);
        sc[1]=new sclass("C002", 3);
        sc[2]=new sclass("C003", 4);
    }
    void initst()
    {   st=new Student[5];                           //学生数组初始化
        st[0]=new Student("S001","zhang shan",'f',78,86,96);
        st[1]=new Student("S002","li si",'m',65,77,89);
        st[2]=new Student("S003","wang wu",'m',98,77,87);
        st[3]=new Student("S004","zhou lin",'f',73,95,93);
        st[4]=new Student("S005","qian dong",'m',87,93,91);
    }

    void accuscore()                                 //计算学分绩
    {   float as=0;                                  //分数和
        int cs=0;                                    //学分绩
        for (int i=0; i<st.length; i++)
```

```java
        { for(int j=0; j<sc.length; j++)
            {   as+=sc[j].credit * st[i].score[j];
                cs+=sc[j].credit;
            }
            st[i].avgscore=(float)(as/cs);
        }
    }
    void sort()                                       //排序
    {   Student tmp=new Student();                    //临时变量
        for(int i=0; i<st.length; i++)
        {   for(int j=i+1; j<st.length; j++)
            if(st[i].avgscore<st[j].avgscore)         //按学分绩排序
            {   tmp=st[i];
                st[i]=st[j];
                st[j]=tmp;
            }
        }
    }
    void print()                                      //输出学生的信息
    {   System.out.println("-----------------------------------");
        System.out.println("the info of the students as follow:");
        System.out.println("-----------------------------------");
        for(int i=0; i<st.length; i++)
        {   st[i].print();
        }
    }
    void print2()                                     //输出排名信息
    {   System.out.println("-----------------------------------");
        System.out.println("sorted by average score, students as follow:");
        System.out.println("-----------------------------------");
        for(int i=0; i<st.length; i++)
        {   st[i].print2();
        }
    }
    public static void main(String[] args)
    {   Stuscore stc=new Stuscore();
        stc.initsc();
        stc.initst();
        stc.accuscore();
        stc.print();
        stc.sort();
        stc.print2();
    }
}
```

程序的执行结果如图 4-6 所示。

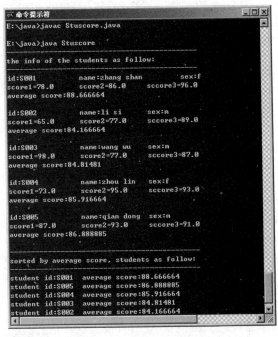

图 4-6　Stuscore 类的执行结果

4.11　使用数组保存书、CD、磁带等的信息,并能实现插入、删除、查找功能。插入、删除时,要显示操作正确与否的信息;查找时按关键字值进行查找,显示查找结果。

解:根据题意,对书、CD、磁带等分别定义相关的类为 Book 类、CD 类及 Tape 类。对这 3 个类分别定义相关的属性,例如对于 Book 类,定义书名、作者、出版社、刊号、出版日期、页数、摘要等;对于 CD 类,可以定义 CD 编号、CD 名、价格、发行公司等;对于 Tape 类,可以定义磁带名、演唱(奏)者、价格、发行公司等。对各个类还可以定义并实现相关的方法,如给各实例赋值、查询某实例相关的属性、对某类实例按某个属性进行排序等。

下面以 CD 为例,定义 CD 类及相关的属性,包括 CD 编号 cdid、CD 名 cdname、价格 price、发行公司 press 共 4 个,并实现了对 CD 的管理。对于 CD 类的管理,定义了 CDManager 类。它继承了 Vector 类,用于实现 CD 实例的插入、删除、查找和打印功能。

程序代码实现如下:

```
import java.*;
import java.io.*;
import java.util.Vector;

class CD
{   String cdid;                        //CD 编号
    String cdname;                      //CD 名
    float price;                        //价格
    String press;                       //发行公司
```

```java
    CD(String cdid, String cdname, float price, String press)    //带参数的构造方法
    {   this.cdid=cdid;
        this.cdname=cdname;
        this.price=price;
        this.press=press;
    }
    CD(){ }                                    //无参数的构造方法

    void setID(String cdid)                    //记录 CD 编号
    {   this.cdid=cdid;}
    void setName(String cdname)                //记录 CD 名
    {   this.cdname=cdname;}
    void setPrice(float price)                 //记录 CD 价格
    {   this.price=price;}
    void setPress(String press)                //记录 CD 的发行公司
    {   this.press=press;}
    void print()                               //输出 CD 信息
    {   System.out.println("ID:"+this.cdid+"\t"+"Name:"+this.cdname+"\t"
            +"Price:"+this.price+"\t"+"Press:"+this.press);
    }
};

public class CDManager extends Vector          //对 CD 的管理
{   Vector cds;

    CDManager()
    {   cds=new Vector(50, 1);
    }

    int addcd(CD tmpcd)                        //添加一盘 CD
    {   cds.addElement(tmpcd);                 //addElement()是 Vector 类的方法
        return 1;
    }
    int deletecd()                             //删除一盘 CD
    {   if(cds.isEmpty()==false)
        {   int i=cds.size();
            cds.removeElementAt(i-1);
            return 1;
        }
        else
        {   return 0;
        }
    }
```

```java
    CD findcd(String cid)                //按 CD 编号查找 CD
    {   CD tmpcd=new CD();
        int i=0;
        for(i=0; i<cds.size(); i++)
        {   CD getcd=(CD)cds.elementAt(i);    //获取每盘 CD
                                              //判定获取的 CD 的编号,并与目标相比较
            if(((getcd.cdid).trim()).compareTo(cid.trim())==0)
            {   tmpcd=(CD)cds.elementAt(i);//相等
            }
        }
        return tmpcd;                     //返回
    }

    public void printcd()                 //输出 CD 信息
    {   CD tmpcd;
        int length=cds.size();            //已有 CD 的数量
        if(length>0)
        {   System.out.println("Number of vector elements is "+length+" and they are:");
            for (int i=0; i<length; i++)  //顺序输出
            {   tmpcd=(CD)cds.elementAt(i);
                System.out.println("ID:"+tmpcd.cdid+"\t"+"Name:"+tmpcd.cdname
                    +"\t"+"Price:"+tmpcd.price+"\t"+"Press:"+tmpcd.press);
            }
        }
        else
        {   System.out.println("There is no one cd ,please add.");
        }
    }

    String inputstr()                     //输入字符串
    {   String pm="";
        try{
            InputStreamReader reader=new InputStreamReader(System.in);
            BufferedReader input=new BufferedReader(reader);
            pm=input.readLine();
        }catch (IOException e)
        {   System.out.println("IO Exception occur...");
        }
        return pm;
    }
    public static void main(String[] args)
    {   CDManager cm=new CDManager();
        int count=1;
        while(count!=0)
```

```java
{   System.out.print("------------------------------------\n ");
    System.out.print("which of following do you want to do:(1,2,3,4,0)\n ");
    System.out.print("1.add cd.\n ");
    System.out.print("2.find cd.\n ");
    System.out.print("3.delete cd.\n ");
    System.out.print("4.print all cds.\n ");
    System.out.print("0.exit.\n ");
    try
    {   count=Integer.parseInt(cm.inputstr());
    }catch(NumberFormatException ne)
    {   System.out.println("invalid data format ...");
    }
    switch (count)                                  //根据输入，进行相应的操作
    {
        case 1:                                     //添加一盘 CD
            CD cd1=new CD();
            System.out.print("please input cd ID : ");
            cd1.setID(cm.inputstr());               //输入并记录编号
            System.out.print("please input cd Name : ");
            cd1.setName(cm.inputstr());             //输入并记录 CD 名
            System.out.print("please input cd Price : ");
            try{
                float pc=Float.parseFloat(cm.inputstr());
                                                    //字符串转浮点数
                cd1.setPrice(pc);                   //得到 CD 的价格
            }catch(NumberFormatException ne)
            {   System.out.println("please input format number!");}
            System.out.print("please input cd Press : ");
            cd1.setPress(cm.inputstr());            //输入并记录发行公司
            int i=cm.addcd(cd1);                    //添加一盘 CD
            if(i>0)                                 //添加成功
            {   System.out.print("insert successful!\n ");
            }
            else                                    //添加不成功
            {   System.out.print("insert failed!\n");
            }
            break;
        case 2:                                     //按 CD 编号查询 CD
            System.out.println("please input ID of cd you want to find: ");
            String id=cm.inputstr();                //输入目标编号
            CD fcd=cm.findcd(id);                   //按编号查找
            if (fcd.cdid==null)                     //查找不成功
            {   System.out.println("There is no cd you want. ");
            }
            else                                    //查找成功
            {   System.out.println("info of cd you want to find is: ");
```

```java
            fcd.print();                    //输入找到的CD信息
        }
        break;
    case 3:                                 //删除一盘CD
        if(cm.deletecd()>0)                 //删除成功
        {   System.out.print("delete successful!\n ");
        }
        else                                //删除失败
        {   System.out.print("delete failed!\n ");
        }
        break;
    case 4:                                 //输出所管理的所有CD的全部信息
        cm.printcd();
        break;
    case 0:                                 //退出
        System.exit(0);
        default:
        break;
    }//end of switch
  }//end of while
 }
}
```

作为示例,上述程序仅实现了对CD的管理。读者可以参照此例,继续实现对书及磁带的管理,也可以实现添加、删除、输出相关信息等操作。在此基础之上,还可以实现查询、排序等功能,以完善管理。

为了测试程序,首先添加几条CD的信息,显示是否添加正确,并显示目前库中的所有CD。接下来按照CD的编号进行查询,查找所需的CD并删除之,以测试删除操作的执行是否正确。添加CD操作的执行结果如图4-7所示。显示所有CD信息的结果如图4-8所示。按CD编号进行查询的结果如图4-9所示。删除CD操作的结果如图4-10所示。所有的操作完毕,显示剩余CD的结果如图4-11所示。

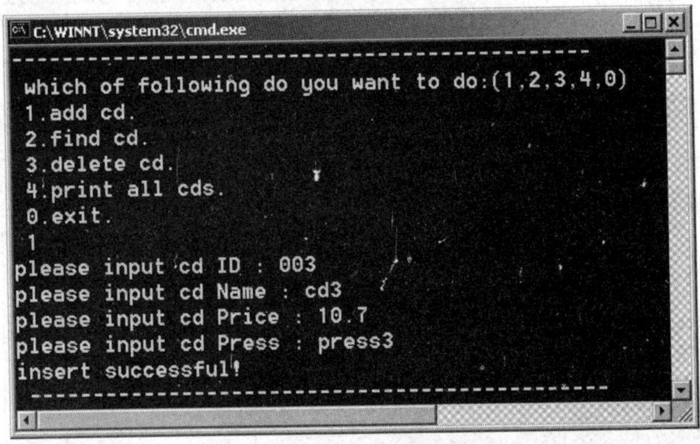

图4-7 添加CD的实现

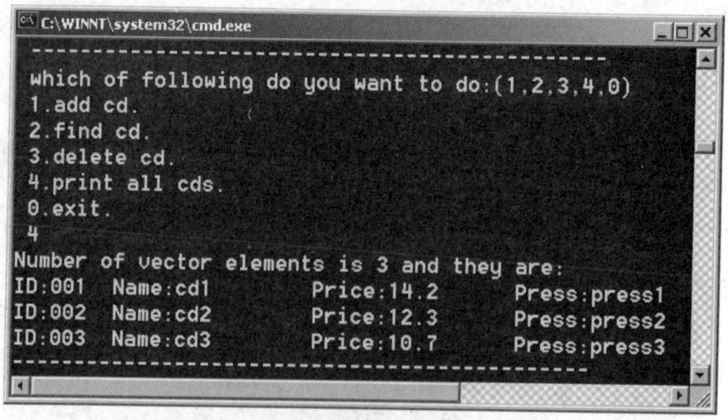

图 4-8 显示目前所有 CD 的信息

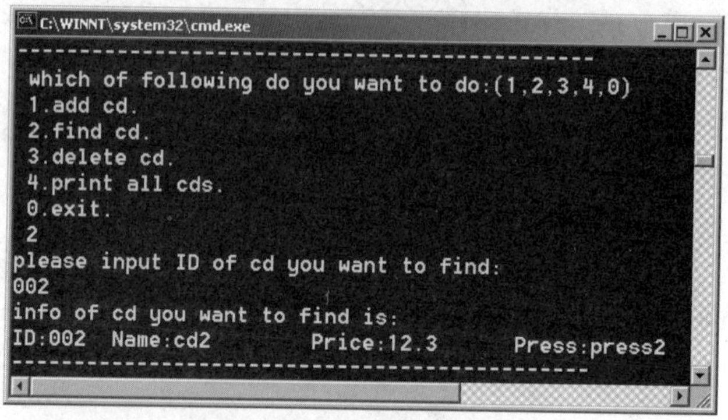

图 4-9 按 ID 查找相应的 CD

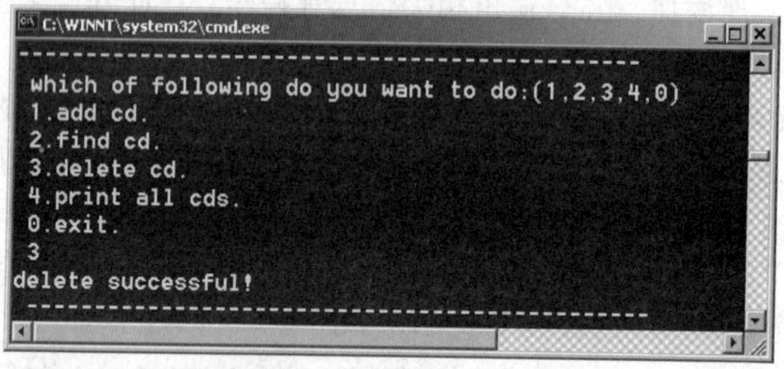

图 4-10 删除一盘 CD

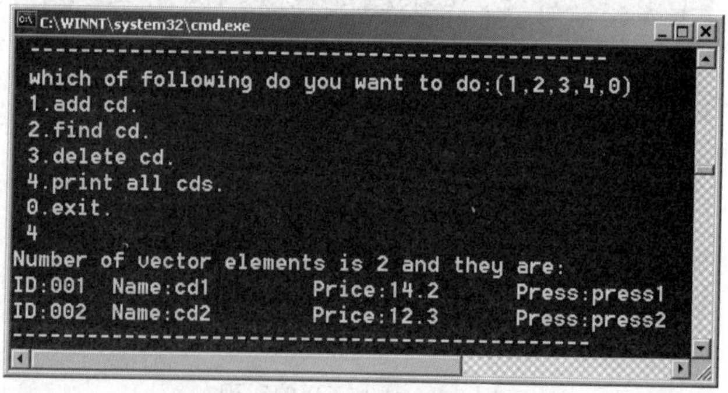

图 4-11 显示目前剩余 CD 的信息

第5章 对象和类的进一步介绍

5.1 详细说明类是如何定义的,解释类的特性及它的几个要素。

解:Java 程序设计就是定义类的过程,Java 程序中的所有代码都包含在类中。类可以看作是数据的集合及操作这些数据所需的方法的整合。

Java 中的类分两种,一种是系统预定义的类,这些类组成 Java 类库。Java 类库是一组由软件供应商编写好的程序模块,完成常用的基本功能和任务,可由程序编写人员直接调用。正是由于有了这些类库,程序员才有了很好的辅助工具,不必将精力浪费在一些简单常见的功能实现上。基本类库提供的这些功能,使得程序员站在了一个较高的起点上,他们可以把主要精力关注在更加复杂的工作上。这些定义好的类根据实现功能的不同,划分成不同的集合,每个集合称为一个包。Sun 公司提供的 JDK 中共有 43 个大包。

除去系统预定义的类之外,还有一种是用户程序自己定义的类,当然这其中又包括其他程序员定义的类和自己定义的类。这些类都显式或隐式地派生于 Java 中某个预定义的类。不论是预定义的类,还是程序员自己定义的类,每个类中一般都包含属性和方法。属性即是数据,属性值表明一个对象的状态;方法决定类有哪些可利用的手段,即可通过哪些函数来操作这些数据。

类的具体格式如下:

```
修饰符 class 类名[extends 父类名]{
    类型 成员变量1;
    类型 成员变量2;
    ……
    修饰符 类型 成员方法1(参数列表){
        类型 局部变量;
        方法体
    }
    修饰符 类型 成员方法2(参数列表){
        类型 局部变量;
        方法体
    }
    ……
}
```

类定义的第一行是类头,关键字 class 表明这里定义的是一个类。class 前的修饰符允许有多个,用来限定所定义的类的使用方式。

类名是用户为该类所起的名字,它应该是一个合法的标识符,并尽量遵从命名约定。

extends 是关键字。如果所定义的类是从某一个父类派生而来,那么,父类的名字应写在 extends 之后。如果不写的话，则隐式表明继承于 Object 类。Java 不允许多重继承，所以如果有父类的话，只能有一个父类。

类头后面的部分称为类体，类体用一对大括号括起来，含有两部分，一部分是数据成员变量，另一部分是成员方法。数据成员变量可以含有多个，这是类的静态属性，表明类的实例目前所处的状态。类的不同实例对应各自不同的属性值，因此有些属性值可用来标识不同的实例。成员变量前面的类型是该变量的类型。成员方法也可以有多个，其前面的类型是方法返回值的类型。方法对应类的行为和操作。方法体是要执行的真正语句。在方法体中还可以定义该方法内使用的局部变量，这些变量只在该方法内有效。

类可以是 public 的，表明任何对象都可以使用或扩展这个类；也可以是 friendly 的，表明它可以被同一个包中的对象使用。类还可以是 final 的，表明它不可以再有子类，与之相对的，使用 abstract 修饰的类必须要有子类。

5.2 给出 3 个类的定义：

```
class ParentClass {}
class SubClass1 extends ParentClass {}
class SubClass2 extends ParentClass {}
```

并分别定义 3 个对象：

```
ParentClass a=new ParentClass();
SubClass1 b=new SubClass1();
SubClass2 c=new SubClass2();
```

若执行下面的语句：

```
a=b;
b=a;
b=(SubClass1)c;
```

会有什么结果？分别从下面的选项中选择正确的答案。

（1）编译时出错。

（2）编译时正确，但执行时出错。

（3）执行时完全正确。

解：3 行类定义分别定义了 3 个类，一个父类 ParentClass 及它的两个子类 SubClass1 和 SubClass2。后面的 3 行则分别为每个类说明了一个实例，其中，a 是 ParentClass 类的实例，b 是 SubClass1 类的实例，c 是 SubClass2 类的实例。由于 SubClass1 和 SubClass2 都是派生于 ParentClass 的子类，所以 b 和 c 也同时是 ParentClass 类的实例。

Java 中允许使用对象之父类类型的一个变量指示该对象，称为转换对象（casting）。关于转换对象的使用，遵从对象引用的赋值兼容原则。所谓对象引用的赋值兼容原则是指允许把子类的实例赋给父类的引用，但不允许把父类的实例赋给子类的引用。实际编程时，可以使用 instanceof 运算符来判明一个引用指向的是哪个类的实例。如果父类的引用指向的是子类实例，就可以转换该引用，恢复对象的全部功能。

本题中,可以进行下面的测试:

```
boolean tagb1=b instanceof ParentClass;
boolean tagc1=c instanceof ParentClass;
```

tagb1 和 tagc1 的值都是 true,表明 b 和 c 是子类实例的同时,也是父类的实例。反过来,父类的实例不是子类的实例,例如下面的测试:

```
boolean taga2=a instanceof SubClass1;
boolean taga3=a instanceof SubClass2;
```

taga2 和 taga3 的值都是 false。b 和 c 是不同子类的实例,所以如下测试:

```
boolean tagb3=b instanceof SubClass2;
boolean tagc2=c instanceof SubClass1;
```

将出现编译错误。下面针对题目中的 3 条语句分别进行测试。

(1) 执行 a=b;时,a 指向父类的实例,b 指向子类的实例。由于是将子类实例赋给父类实例,因此编译及执行都是正确的。该语句将执行子类中的方法,如果子类中没有重写父类中的方法,则将执行父类中的方法。例如下面的程序中,父类和子类中都定义了 value 成员和 getValue()方法,将子类的实例赋给父类引用后,此时 a 的值是子类的实例,再执行语句:

```
taga2=a instanceof SubClass1;
```

则 taga2 的值应为 true。在给 a 分配的内存中既包括子类中 value 的值,也含有父类中 value 的值。调用 a.getValue()方法时,先在子类中查找这个方法是否存在,如果有,则返回子类中 value 的值 1;若没有,则查找父类中的同名方法,并返回父类中的值 0。

完整的测试代码如下:

```java
import java.util.*;

class ParentClass
{   public ParentClass()                //父类的构造方法
    {   value=0;
    }
    public int getValue()               //父类的求值方法,返回父类中 value 的值
    {   return value;
    }
    public void setValue(int y)         //给父类的属性 value 赋值
    {   value=y;
    }

    private int value;
}
class SubClass1 extends ParentClass
```

```java
{   public SubClass1()              //子类1的构造方法,value值为1
    {   value=1;
    }
    public int getValue()           //子类的同名求值方法,返回子类1中value的值
    {   return value;
    }
    public int getClassValue1()     //子类的特殊求值方法,返回classvalue1的值
    {   return classvalue1;
    }
    private int value;
    private int classvalue1=11;
}
class SubClass2 extends ParentClass
{   public SubClass2()              //子类2的构造方法,value值为2
    {   value=2;
    }
    public int getValue()           //子类的同名求值方法,返回子类2中value的值
    {   return value;
    }
    public int getClassValue2()     //子类的特殊求值方法,返回classvalue2的值
    {   return classvalue2;
    }
    private int value;
    private int classvalue2=22;
}

public class Test2
{   public static void main(String[] args)
    {   ParentClass a=new ParentClass ();              //父类实例
        SubClass1 b=new SubClass1();                    //子类1的实例
        SubClass2 c=new SubClass2();                    //子类2的实例

        a=b;                                            //a指向子类1的实例
        System.out.println("a="+a.getValue());          //返回子类1中的属性值
    }
}
```

程序的执行结果如下:

a=1

(2) 执行b=a;时,由于是将父类的实例赋给子类的变量,因此会出现编译错误,错误类型是变量类型不匹配。

(3) 执行b=(SubClass1)c;时,由于b和c是不同类的实例,因此也会出现编译错误,错误类型是变量类型不能转换。

【拓展思考】

(1) new 运算符执行什么动作?

解：new 运算符创建指定类的一个新实例(对象)。然后调用类的构造方法设置新生成的对象。

(2) 什么是 null 引用?

解：null 引用是不指向任何对象的引用。用保留字 null 来检查空引用,以避免对空引用的访问。

(3) 什么是别名?

解：如果两个引用指向同一个对象,则它们互为别名。通过一个引用改变对象的状态,也就改变了另一个引用指向的对象,因为实际上只有一个对象。仅当一个对象再没有引用指向它时,才被垃圾收集所标记。

5.3 什么是抽象类? 它如何定义? 下面的几个定义哪些是正确的?

(1)

```
class alarmclock {
    abstract void alarm();
}
```

(2)

```
abstract alarmclock {
    abstract void alarm();
}
```

(3)

```
class abstract alarmclock {
    abstract void alarm();
}
```

(4)

```
abstract class alarmclock {
    abstract void alarm();
}
```

(5)

```
abstract class alarmclock {
    abstract void alarm(){
        System.out.println("alarm!")
    };
}
```

解：如果一个方法只有方法的声明,而没有方法的实现,则称为抽象方法(abstract method)。含有抽象方法的类通常称为抽象类(abstract class)。在 Java 中可以通过关键字 abstract 把一个类定义为抽象类,每一个未被定义具体实现的抽象方法也应标记为 abstract。抽象类是表示一般概念的类。在抽象类中可定义公共特征及方法签名,然后由其子类来继承它们。

在抽象类中可以包括被它的所有子类共享的公共行为,也包括被它的所有子类共享的公共属性。因为抽象类中含有抽象方法,所以不能用抽象类作为模板来创建对象,必须生成抽象类的一个非抽象的子类后才能创建实例。这是因为一个实例的任何方法都必须已经具体实现了。抽象类可以包含常规类能够包含的任何元素,当然也包括构造方法,因为子类可能需要继承这种方法。抽象类中当然包含抽象方法,这种方法只有方法的声明,而没有方法的实现。这些方法将在抽象类的子类中被具体实现。只有实现了所有抽象方法的子类才能创建对应的实例。除了抽象方法,抽象类中当然也可以包含非抽象方法,反之,不能在非抽象的类中声明抽象方法。也就是说,只有抽象类才能具有抽象方法。

根据抽象方法和抽象类的定义和规则,再来分析题目中的 5 个语句说明。第 1 个示例中,方法 alarm()定义为抽象方法,那么方法所在的类也必须声明为抽象类型;第 2 个示例中,虽然方法和类都说明是抽象的,但缺少关键字 class;第 3 个示例中,各关键字的次序不正确,abstract 关键字要放在 class 之前;第 5 个示例中,虽然方法 alarm()说明为抽象方法,但方法体不为空,这显然有矛盾。只有第 4 个示例的说明才是正确的。

【拓展思考】

父类的所有成员都能被子类继承吗?请解释理由。

解:如果类成员是私有可见的,则不能继承它们,就是说,在子类中不能通过名字来直接引用它们。但对子类来说,这样的成员确实存在,可以间接引用。

5.4 什么叫作方法重载?什么叫作方法重写?它们之间的区别是什么?

解:同一个类中,可以定义同名的多个方法。它们的不同之处在于参数列表不同,这其中包括参数的个数不同或是对应的参数类型不完全相同。这就是方法的重载(overload)。当对该类的实例调用相应的方法时,系统将依据所带参数的个数及类型从同名的多个方法中来选择参数列表满足要求的方法。不只如此,在不同的类中也可以定义有相同方法名的方法。

使用类的继承关系,可以从已有的类产生一个新类,在原有特性基础上,增加新的特性。原类及新类分别称为父类及子类。如果父类中原有的方法不能满足子类的要求,可以在子类中对父类的方法重新编写代码。这称为方法重写(override),也称为方法的隐藏,意思是子类中看不到父类中的实现代码。子类中定义的方法所用的名字、返回类型及参数表和父类中方法使用的完全一样,从逻辑上看就是子类中的成员方法隐藏了父类中的同名方法。

从对方法重载和方法重写的分析可以看出,它们之间的区别主要有:

- 方法重载时参数列表必须是不同的,使系统能区别出到底调用哪个方法,而方法重写时参数列表可以是相同的。
- 方法重载时,方法的返回类型可能不同,也可能相同;而方法重写时,子类中方法的返回类型和父类中同名方法的返回类型完全一样。
- 方法重载多出现在同一个类中,方法重写必须是在父子类中。

【拓展思考】

(1) 什么是多态？

解：多态是引用变量在不同时刻指向不同类型对象的一种能力。通过这样的引用，调用的方法可以在不同的时刻，根据对象引用的类型与不同的方法进行绑定。

(2) 继承如何支持多态？

解：在 Java 中，使用父类声明的引用变量可以指向子类的对象。两个类包含有相同签名的方法时，父类引用是多态的。

(3) 与多态相关的重写是如何实现的？

解：当子类重写父类方法的定义时，这个方法就有了两个版本。用多态引用调用这个方法时，调用的方法版本取决于所用对象的类型，不取决于引用变量的类型。

(4) 如何使用接口完成多态？

解：接口名可作为引用类型使用。这样的引用变量可指向实现该接口的任一类的任一对象。因为所有的类实现同一个接口，它们有公共的签名，因此可以动态绑定。

(5) 单重继承和多重继承之间的差别是什么？

解：对于单重继承，只能从一个父类派生子类；而对于多重继承，一个类可从多个父类中派生，并继承每个父类的特性。多重继承的问题是，必须解决当两个或多个父类中有同名的属性或方法时引起的冲突。Java 只支持单重继承。

5.5 什么是 null 引用？

解：null 引用是不指向任何对象的引用。用保留字 null 来检查空引用，以避免对空引用的访问。

在 Java 中，当执行 new 为一个对象分配内存时，Java 自动初始化所分配的内存空间。对于引用，即对象类型的任何变量，使用一个特殊的值 null 进行初始化，它表示引用不指向任何对象。运行过程中系统发现使用了这样一个引用时，可以立即停止进一步的访问，不会给系统带来任何危险。

5.6 关键字 this 和关键字 super 在成员方法中的特殊作用是什么？

解：在 Java 中，this 引用总是指向当前对象，即 this 引用所在的对象。如果在类的成员方法中访问类的成员变量，可以使用关键字 this 指明要操作的对象。在构造方法中，this 还有一个用法，即作为构造方法的第一个语句，它的形式是 this(参数表)，这个构造方法就会调用同一个类的另一个构造方法。

如果子类已经重写了父类中的方法，但在子类中还想使用父类中被隐藏的方法，或者子类中定义了和父类中同名的成员变量，还想使用父类中隐藏的成员变量，可以使用 super 关键字。

5.7 仿照书中的例子，构造一个类，并使其具有多个相互调用的构造方法；然后构造它的子类，在构造方法中利用关键字 super 来调用父类的构造方法。

解：在习题 2.10 中，定义了教师类及其子类。父类及子类中有多个相同的属性，对这些属性的访问方法也存在于父类及子类中。在习题 2.10 的实现中，父子类中的构造方法是独立的，它们之间没有关系。实际上，由于父子类中有多个相同的属性，子类的构造方法就可以借用父类的构造方法，以简化代码。除构造方法外，子类的任何成员方法都可以借用父类同名的成员方法,并在此基础上,增加自己的特殊处理部分。

修改习题 2.10 的实现,包括修改其中定义的类及成员方法。先定义一个类 SchoolTeacher,其中包括 3 个互相调用的构造方法。在此基础上派生子类 ResearchSchoolTeacher,其构造方法中使用 super 来调用父类的构造方法,对父类及子类共有的属性进行赋值。除了在构造方法中可以使用 super 调用父类的构造方法外,还可以在一般的方法中也调用父类的方法。例如本题中,当输出 ResearchSchoolTeacher 类实例的信息时,因前 6 个属性是与父类实例相同的属性,所以可以借用父类中的输出方法:

```
super.print();                                    //使用父类的输出方法
```

这个语句调用父类的输出方法,先输出前 6 个属性的值,然后再使用下面的两条语句:

```
System.out.print("The SchoolTeacher research field is:  ");
System.out.println(this.getResField());           //特殊属性的输出
```

完成对该类实例 resField 属性的输出。

程序代码实现如下:

```java
import java.lang.*;

class Date                                        //定义日期类
{   int day;
    int month;
    int year;
    Date (int day, int month, int year)
    {   this.day=day;
        this.month=month;
        this.year=year;
    }
    Date ()
    {   this.day=8;
        this.month=11;
        this.year=2012;
    }
    public int getYear()                          //返回年
    {   return year;
    }
    public int getMonth()                         //返回月
    {   return month;
    }
    public int getDay()                           //返回日
    {   return day;
    }
    public void setDate(Date SpeDate)             //设置日期
```

```java
        {   year=SpeDate.getYear();
            month=SpeDate.getMonth();
            day=SpeDate.getDay();
        }
    }

public class SchoolTeacher                          //定义教师类,这是基类
{   private String name;                            //教师名字
    private boolean sex;                            //性别,true 表示男性;false 表示女性
    private Date birth;                             //教师出生日期
    private String salaryID;                        //教师工资号
    private String depart;                          //教师所属系
    private String posit;                           //教师职称

    String getName()                                //返回教师名字
    {   return name;
    }
    void setName (String name)                      //记录教师名字
    {   this.name=name;
    }
    boolean getSex()                                //返回教师性别
    {   return sex;
    }
    void setSex (boolean sex)                       //记录教师性别
    {   this.sex=sex;
    }
    Date getBirth()                                 //返回教师出生日期
    {   return birth;
    }
    void setBirth (Date birth)                      //记录教师出生日期
    {   this.birth=birth;
    }
    String getSalaryID()                            //返回教师工资号
    {   return salaryID;
    }
    void setSalaryID (String salaryID)              //记录教师工资号
    {   this.salaryID=salaryID;
    }
    String getDepart()                              //返回教师所属系所名
    {   return depart;
    }
    void setDepart (String depart)                  //记录教师所属系所名
    {   this.depart=depart;
    }
```

```java
    String getPosit()                              //返回教师职称
    {   return posit;
    }
    void setPosit(String posit)                    //记录教师职称
    {   this.posit=posit;
    }

    public SchoolTeacher(String name)              //只含一个属性参数的构造方法
    {   this.name=name;
    }

    //含有 3 个属性参数的构造方法
    public SchoolTeacher(String name, boolean sex, Date birth)
    {   this(name);                                //调用只含一个参数的构造方法
        this.sex=sex;                              //对其余的两个属性赋值
        this.birth=birth;
    }

    //含有全部 6 个属性参数的构造方法
    public SchoolTeacher(String name, boolean sex, Date birth,
        String salaryid, String depart, String posit)
    {   this(name, sex, birth);                    //调用含 3 个参数的构造方法
        this.salaryID=salaryid;                    //对其余的 3 个属性赋值
        this.depart=depart;
        this.posit=posit;
    }
    public void print()                            //输出教师基本信息
    {   System.out.print("The SchoolTeacher name:   ");
        System.out.println(this.getName());
        System.out.print("The SchoolTeacher sex:   ");
        if (this.getSex()==false)
        {   System.out.println("女");
        }
        else
        {   System.out.println("男");
        }
        System.out.print("The SchoolTeacher birth:   ");
        System.out.println(this.getBirth().year +"-"+this.getBirth().month
            +"-"+this.getBirth().day );
        System.out.print("The SchoolTeacher salaryid:   ");
        System.out.println(this.getSalaryID());
        System.out.print("The SchoolTeacher posit:   ");
        System.out.println(this.getPosit());
        System.out.print("The SchoolTeacher depart:   ");
```

```java
        System.out.println(this.getDepart());
    }
    public static void main (String [] args)
    {   Date dt1=new Date(12, 2, 1985);           //创建日期实例,作为教师的出生日期
        Date dt2=new Date(2, 6, 1975);
        Date dt3=new Date(11, 8, 1964);
        Date dt4=new Date(10, 4, 1975);
        Date dt5=new Date(8, 9, 1969);

        //创建两个教师实例,一个为父类的实例,另一个为子类的实例
        SchoolTeacher t1=new SchoolTeacher("zhangsan", false,dt1, "123",
            "CS", "Prefessor");
        ResearchSchoolTeacher rt=new ResearchSchoolTeacher("lisi", true, dt2, "421",
            "software engineering", "associate professor", "Software");

        //分别调用各自的输出方法,输出相应的信息
        System.out.println("-----------------------------------");
        t1.print();                               //输出普通教师的信息
        System.out.println("-----------------------------------");
        rt.print();                               //输出研究系列教师的信息
        System.out.println("-----------------------------------");
    }
}

class ResearchSchoolTeacher extends SchoolTeacher    //研究系列教师类的定义
{   private String resField;                         //增加的研究领域属性
    String getResField()                             //返回研究领域属性
    {    return resField;
    }
    void setResField (String resField)               //记录研究领域属性
    {    this.resField=resField;
    }
    public ResearchSchoolTeacher(String name, boolean sex, Date birth,
        String salaryid, String depart, String posit, String resField)
    {   //使用父类的构造方法,对共有的 6 个属性进行赋值
        super(name, sex, birth, salaryid, depart, posit);
        this.resField=resField;                      //特殊属性的赋值
    }
    public void print()
    {   System.out.println("One of Research SchoolTeachers' info is ");
        super.print();                               //使用父类的输出方法
        System.out.print("The SchoolTeacher research field is:  ");
        System.out.println(this.getResField());      //特殊属性的输出
```

        }
    }

程序的执行结果如图 5-1 所示。

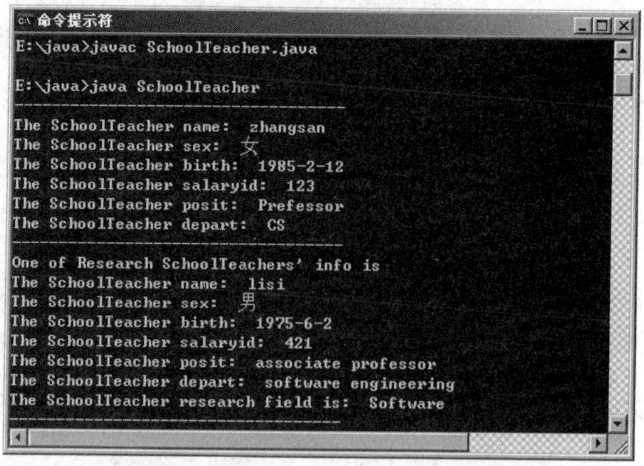

图 5-1 SchoolTeacher 类的执行结果

5.8 什么是静态方法和静态变量,它们同普通的成员方法和成员变量之间有何区别?

解:一般来讲,一个类的多个实例中各变量的值可以各不相同,但有时需要让一个类的多个实例共享一个变量,也就是对任何一个实例来讲,这个属性的值都是一样的。这样的变量称为类变量,也叫静态变量。可以说静态变量被类的多个实例所共享。

静态变量的作用主要有两个,一是可以实现多个对象之间的通信。当想从一个实例中传值给另一个实例时,可以将这个值放到静态变量中,则另一个实例通过访问静态变量就会很方便地得到这个值。第二个作用是可以用来记录已被创建的对象的个数。有些应用中可能会要求记录一个类的实例个数,当创建完成时,可以将静态变量的值加 1,以完成计数。

静态变量是区别于成员变量或实例变量的,将一个变量定义为静态变量的方法就是将这个变量标记上关键字 static。

如果需要用到在尚未创建一个对象实例的时候就去引用它的程序代码,那么标记上关键字 static 即可实现。这样的方法称为类方法(或称静态方法)。

静态变量和静态方法与普通的成员变量和成员方法间的区别主要体现在以下几方面。

静态方法和静态变量都需要使用 static 来标记,而成员方法和成员变量却不需要这样标记,这是形式上的不同。

类的每个对象具有它自己的、在类中定义的所有实例变量的副本,但对于静态变量,它们共享一个副本。也就是说,每个实例中,静态变量的值都一样,而成员变量的值可以互不相同。这是逻辑上的不同。

静态变量的生命期和实例变量的生命期也是不同的,每次实例化一个对象时,都将创建实例变量,当对象被垃圾收集器销毁时,实例变量也被销毁。就是说,成员变量的生命期与实例相同。相反,当类被第一次装载到内存中时,将创建静态变量,并且静态变量一直存在,直到程序执行结束。静态变量的生命期与类相同。

在使用上,成员方法与静态方法的形式也不一样。静态方法可以通过一个指向类名的引用来访问,即不需实例化。但成员方法必须是创建了实例后,才能访问这个实例的成员方法,并且要限定是通过哪个实例来访问。

5.9 什么是抽象类?什么是抽象方法?它们有什么特点和用处?

解:参看习题 5.3。

如果一个方法只有方法的声明,而没有方法的实现,则称为抽象方法(abstract method)。含有抽象方法的类通常称为抽象类(abstract class)。在 Java 中可以通过关键字 abstract 把一个类定义为抽象类,每一个未被定义具体实现的抽象方法也应标记为 abstract。

从形式上看,在 Java 中,类 class 之前可以使用多种修饰符,用来修饰限定所定义的类的特性。凡是用 abstract 修饰符修饰的类都是抽象类。从实现上来看,抽象类就是没有具体对象的概念类。例如,我们可以定义一个类叫作房子,这是一个抽象概念,我们只知道它应该有门、窗、四壁、屋顶及一些必备设施,但现实生活中我们看到的都是具体的房子,谁也没见过一个抽象的房子。房子可以派生出多个具体的类,就是我们日常所见的具体的楼房、平房、别墅、草房等。

抽象类的定义如下:

```
abstruct class 类名
{    类体;
}
```

抽象类一定要派生子类,父类中的抽象方法可以在子类中具体实现,也可以在子类中继续说明为抽象的。换句话说,抽象类是至少含有一个不完整方法的类。由于抽象类的特殊性,因此不能创建抽象类的实例。

抽象类通常位于面向对象的类层次结构的顶层,它定义了该类的所有子类的对象可能具有的大致行为类型。具体的行为则在各子类中完成。

5.10 什么是终极类、终极方法和终极变量?定义终极类型的目的是什么?

解:和抽象类一般处于层次中的顶层并一定要派生子类相反,另一类处于最底层的类则要求不能再被继承,它们被标记为 final,这样的类称之为终极类(final class),其声明的格式为:

```
final class 类名{
    ……
}
```

成员方法也可被标记为 final 从而成为终极方法(final method),被标记 final 的方法将不能被重写。如果一个变量被标记为 final,则会使它成为一个常量。定义终极类型的目的首先是出于安全考虑,被定义为 final 类型的类不能被继承,被定义为 final 类型的方法和变量调用时不能被修改,这样就确保了按照原始的、正确的形式被使用。

另外,如果某个类的结构和功能已经很完整,不需要生成它的子类,这时可以将其定义为终极类。对标有 final 的方法,系统可以进行相应的优化。编译器编译此类方法时对所生成的代码进行了特殊处理,允许对该方法直接调用,而不再像对待一般成员方法那样使用通常的虚拟调用,即在执行时再决定究竟调用哪个方法,从而提高了编译运行效率。

5.11 什么是接口?接口的作用是什么?它与抽象类有何区别?

解:Java 中不允许有多重继承,即如果一个类有父类的话,它只能有一个父类。这是为了避免在子类中对多父类中同名方法的混乱调用而强制规定的。但 Java 并没有限制从多个类中分别继承特性的自由和方便,因此提供了接口这种机制。

接口(interface)是类的另一种表现方式,它是一种特殊的"类"。更确切地说,它是抽象类功能的另一种实现方法,可将其想象为一个"纯"的抽象类。它也有一个类的基本形式,包括方法名、自变量列表以及返回类型,但不规定方法主体。因此在接口中所有的方法都是抽象方法,都没有方法体。

因为没有方法体,所以接口必须被继承。在实现接口的类中必须实现那些抽象方法,给出方法具体的实现细节。实现接口的这些类的实例对应的是一种代码实现,即使有同名同参数列表的方法也不会再产生混乱。

接口的作用是为了保证多重继承,它可以定义多个类的共同属性。而且,Java 通过允许一个类实现多个接口从而实现了比多重继承更加强大的能力,并具有更加清晰的结构。

接口与抽象类的区别是:

(1) 在接口中所有的方法都是抽象方法,都没有方法体,而在抽象类中定义的方法可以不限于抽象方法。

(2) 接口中定义的成员变量都默认为终极类变量,即系统会将其自动增加 final 和 static 这两个关键字,并且对该变量必须设置初值,而抽象类中没有此限制。

(3) 一个类只能由唯一的一个类继承而来,但可以实现多个接口。

【拓展思考】

(1) 类与接口有什么不同?

解:类可被实例化;接口不能被实例化。接口可以只包含抽象方法和常量。类提供了接口的实现。

(2) 类层次和接口层次之间是如何交叉的?

解:类层次和接口层次不交叉。类可用来派生一个新类,接口可用来派生一个新接口,但两个层次不重叠。

(3) 描述 Comparable 接口。

解:Comparable 接口只包含一个方法 compareTo,它根据所执行对象是小于、等于还是大于要比较的对象,分别返回小于 0、等于 0 或大于 0 的一个整数。

(4) 如何使用接口完成多态?

解:接口名可作为引用类型使用。这样的引用变量可指向实现该接口的任一类的任一对象。因为所有的类实现同一个接口,它们有公共的签名,可以动态绑定。

5.12 什么是 Java 包？以一个你熟悉的包为例，列出其中常用的类及接口。

解：Java 中的包一般均包含相关的类，使用包的目的就是要将相关的源代码文件组织在一起，因为不同的包中的类名可以相同，从而应尽最大可能避免名字冲突。从另一个角度来看，这种机制提供了包一级的封装及存取权限。

以 java.lang 包为例，其中常用的类列在表 5-1 中。

表 5-1 java.lang 包中的常用类

类	功 能
Boolean	Boolean 类将基本类型为 boolean 的值包装在一个对象中
Byte	Byte 类将基本类型 byte 的值包装在一个对象中
Character	Character 类在对象中包装一个基本类型 char 的值
Character.Subset	此类的实例表示 Unicode 字符集的特定子集
Character.UnicodeBlock	表示 Unicode 规范中字符块的一系列字符子集
Class<T>	Class 类的实例表示正在运行的 Java 应用程序中的类和接口
ClassLoader	类加载器是负责加载类的对象
Compiler	Compiler 类主要支持 Java 到本机代码的编译器及相关服务
Double	Double 类在对象中包装了一个基本类型 double 的值
Enum<E extends Enum<E>>	这是所有 Java 语言枚举类型的公共基本类
Float	Float 类在对象中包装了一个 float 基本类型的值
InheritableThreadLocal<T>	该类扩展了 ThreadLocal，为子线程提供从父线程那里继承的值：在创建子线程时，子线程会接收所有可继承的线程局部变量的初始值，以获得父线程所具有的值
Integer	Integer 类在对象中包装了一个基本类型 int 的值
Long	Long 类在对象中包装了基本类型 long 的值
Math	Math 类包含基本的数字操作，如指数、对数、平方根和三角函数
Number	抽象类 Number 是 BigDecimal、BigInteger、Byte、Double、Float、Integer、Long 和 Short 类的超类
Object	类 Object 是类层次结构的根类
Package	Package 对象包含有关 Java 包的实现和规范的版本信息
Process	ProcessBuilder.start() 和 Runtime.exec 方法创建一个本机进程，并返回 Process 子类的一个实例，该实例可用来控制进程并获取相关信息
ProcessBuilder	此类用于创建操作系统进程
Runtime	每个 Java 应用程序都有一个 Runtime 类实例，使应用程序能够与其运行的环境相连接

续表

类	功 能
RuntimePermission	该类用于运行时权限
SecurityManager	安全管理器是一个允许应用程序实现安全策略的类
Short	Short 类在对象中包装基本类型 short 的值
StackTraceElement	堆栈跟踪中的元素，它由 Throwable.getStackTrace()返回
StrictMath	类 StrictMath 包含了用于执行基本数字运算的方法，如基本指数、对数、平方根和三角函数
String	String 类代表字符串
StringBuffer	线程安全的可变字符序列
StringBuilder	一个可变的字符序列
System	System 类包含一些有用的类字段和方法
Thread	线程是程序中的执行线程
ThreadGroup	线程组表示一个线程的集合
ThreadLocal<T>	该类提供了线程局部变量
Throwable	Throwable 类是 Java 语言中所有错误或异常的超类
Void	Void 类是一个不可实例化的占位符类，它保持一个对代表 Java 关键字 void 的 Class 对象的引用

常用的接口列在表 5-2 中。

表 5-2 java.lang 包中的常用接口

接 口	功 能
Appendable	能够被追加 char 序列和值的对象
CharSequence	CharSequence 是 char 值的一个可读序列
Cloneable	此类实现了 Cloneable 接口，以指示 Object.clone()方法可以合法地对该类实例进行按字段复制
Comparable<T>	此接口强行对实现它的每个类的对象进行整体排序
Iterable<T>	实现这个接口允许对象成为"foreach"语句的目标
Readable	Readable 是一个字符源
Runnable	Runnable 接口应该由那些打算通过某一线程执行其实例的类来实现
Thread..UncaughtExceptionHandler	当 Thread 因未捕获的异常而突然终止时，调用处理程序的接口

5.13 new 操作符完成哪些功能？

解：new 操作符可以为新对象分配空间，也就是要调用构造函数，创建对象的实例。

5.14 什么是变量声明?

解:Java 语言中,通过在对象类型后紧跟一个变量名来对变量进行声明。例如:int num。变量声明也叫变量说明,根据变量是简单变量还是复合类型变量,其说明的效果也分两种结果。对于简单变量,声明即为创建,也就是在内存中分配了适当的存储空间,并完成了初始化,变量声明后可以直接使用。而对于复合类型变量,声明只是定义了对所声明变量的一个引用,这个引用没有指向任何可用的存储单元,还需要通过 new 运算符来实例化该对象,也就是分配内存,并让引用指向这个存储区的首地址。

5.15 Java 中访问控制权限分为几种?它们所对应的表示关键字分别是什么?意义如何?

解:限定访问权限的修饰符有公有(public)、私有(private)、受保护(protected)和友元(friendly)。这些修饰符既可以用来修饰类,又可以用来修饰类中的成分,它们决定所修饰成分在程序运行时被处理的方式。

- public

用 public 修饰的成分表示是公有的,也就是它可以被其他任何对象访问。

- private

和它的名字"私有"一样,类中限定为 private 的成员只能被这个类本身访问,在类外不可见。

- protected

用该关键字修饰的成分是受保护的,只可以被同一包内的类及其子类的实例对象访问。

这 3 个限定符不是必须写的,如果不写,则表明是"friendly",相应的成分可以被所在包中的各类访问。访问权限修饰符列在表 5-3 中。

表 5-3 访问权限修饰符

类 型	无修饰符	private	protected	public
同一类	是	是	是	是
同一包中的子类	是	否	是	是
同一包中的非子类	是	否	是	是
不同包中的子类	否	否	是	是
不同包中的非子类	否	否	否	是

5.16 为什么有时会用到"-deprecation"参数?利用此参数进行编译所罗列出的方法应做何处理?

解:同其他语言一样,JDK 也在一直不断地发展和完善。从 Sun 公司推出 JDK 第一个正式版本 JDK 1.0 到目前最新的 JDK 5.0 版本,提供给用户的 API 函数有很大的改变。从版本之间的变化而言,从 JDK 8 到 JDK 1.1 之间的变化最为明显,以至于 JDK 1.0 中的某些思想和方法对于学习 Java 语言的人会起到误导作用。

在目前的 JDK 的各个新版本中,被更新的方法和新方法同时存在,来达到版本的向下兼容。但是,在把程序代码由 JDK 1.0 移植到 JDK 新版本时,甚至仅仅是使用原先在

JDK 1.0中可以成功运行的代码在JDK新版本中运行时,都最好使用JDK编译器javac中的"－deprecation"参数来对代码进行编译。

"－deprecation"参数的作用是记录类中使用的被更新的所有方法。也就是说,使用此参数进行编译所罗列出的方法在当前版本中已经升级了,建议查看API文档,改用最新的方法重新编写程序。

5.17 什么叫作内部类？采用内部类的好处是什么？

解：内部类(Inner class),也称嵌套类,是JDK 1.1区别于JDK 1.0版本的一个非常重要的新特性。在JDK最早的版本中,只允许使用顶层类——Java文件只能包含已被声明的包中的成员类。现在JDK 1.1及以上版本中都支持以一个类作为其他类的成员,既可以在语句块中局部定义也可以在表达式中匿名定义。

应用内部类的好处在于它是创建事件接收者的一个方便的方法。

5.18 重新设计习题4.11。设计一个媒体类,其中包含:书、CD及磁带3个子类。按照类的设计模式,完成它们的插入、删除及查找功能。

解：习题4.11中,重点讨论了CD类的属性定义及管理的实现,实现了简单的添加、删除、查询功能。代码中定义了两个类：CD类和CDManager类,使用Vector类实现了CD实例的添加、删除、查找和打印功能。

本例中,将扩充习题4.11的功能。因为是处理多种媒体出版物,所以设计了通用的媒体类MyMedia,包括属性：媒体物品名称mediaName、价格price、出版社(或发行商)press、作者(或艺人)artist,并定义了各自的赋值函数。针对书、CD及磁带,又派生了3个子类。对应于书的子类是MyBook,对应于CD的子类是MyCD,对应于磁带的子类是MyTape。对于各个子类,在继承父类的基础上,又有各自的属性和方法。其中MyBook(书)子类的属性包括ISBN、编辑、出版日期等,MyCD(CD)子类的属性包括ISRC和发行商,MyTape(磁带)子类的属性包括ISRC。本例还对各类的特殊属性实现了基本的打印输出功能。

针对媒体类的访问,设计了MyMediaManager类,可以实现添加、删除和查找功能。其中,查找则按照媒体的价格进行查找。给出价格的范围,找出该价格范围内的媒体。

本例实现了两种存储方式,一是使用数组作为存储结构,二是使用向量机制来处理。

在数组模式下,使用3个数组分别保存3类媒体的信息。另外,删除功能也是按照媒体在数组中的位置进行删除。

使用数组作为存储结构的实现如下：

```
import java.io.*;
import java.util.Date;
import java.text.*;

//媒体父类
class MyMedia
{    String mediaName;                    //物品名称
```

```java
    float price;                                    //物品价格
    String press;                                   //出版社(磁带发行商)
    String artist;                                  //作者(艺人)

    //书名(CD或磁带)名字输入成员函数
    void mediaNameInput()
    {   try
        {   InputStreamReader ir=new InputStreamReader(System.in);
            BufferedReader in=new BufferedReader(ir);
            //判断输入物品名称是否为空
            boolean b=true;
            L1:
                while(b)
                {   mediaName=in.readLine();
                    if(mediaName.matches(""))
                    {   System.out.println("名称不能为空!请重试:");
                        continue L1;
                    }
                    b=false;
                }
        }catch (Exception e) {
            System.out.print(e);
        }
    }

    //价格输入成员函数
    void mediaPriceInput()
    {   try
        {   String p=new String();
            InputStreamReader ir=new InputStreamReader(System.in);
            BufferedReader in=new BufferedReader(ir);
            //判断输入价格是否为非负数
            boolean b=true;
            L2:
                while(b)
                {   p=in.readLine();
                    try
                    {   price=Float.parseFloat(p);
                        if(price<=0) {
                            System.out.print("价格不能是负数!请重试:");
                            continue L2;
                        }
                        b=false;
                    }catch(NumberFormatException e)
```

```java
                    {   System.out.print("非数字!请重试:");
                        continue L2;
                    }
                }
            }catch (Exception e)
            {   System.out.print(e);
            }
        }

        //出版社(磁带发行商)输入成员函数
        void mediaPressInput ()
        {   try
            {   String mp=new String();
                InputStreamReader ir=new InputStreamReader(System.in);
                BufferedReader in=new BufferedReader(ir);
                mp=in.readLine();
                press=mp;
            }catch (Exception e)
            {   System.out.print(e);
            }
        }

        //作者(艺人)输入成员函数
        void artistInput ()
        {   try
            {   InputStreamReader ir=new InputStreamReader(System.in);
                BufferedReader in=new BufferedReader(ir);
                String ma=new String();
                ma=in.readLine();
                artist=ma;
            }catch (Exception e)
            {   System.out.print(e);
            }
        }
    } //end of class MyMedia

    //书子类
    class MyBook extends MyMedia
    {   String editor;                              //书的责任编辑
        Date publishDate=null;                      //出版日期
        String bookISBN;                            //书的ISBN号

        //构造函数
        MyBook(String bn, float bp, String bpr, String ba, String bi, String be)
```

```java
{   mediaName=bn;
    price=bp;
    press=bpr;
    artist=ba;
    bookISBN=bi;
    editor=be;
}

//书相关信息的输入成员函数
public void bookOtherInfo()
{   try
    {   String boi=new String();
        String boe=new String();
        String bpds=new String();
        DateFormat bpddf=DateFormat.getDateInstance();
        InputStreamReader ir=new InputStreamReader(System.in);
        BufferedReader in=new BufferedReader(ir);
        System.out.print("请输入书的ISBN:");
        boi=in.readLine();
        System.out.print("请输入书的编辑:");
        boe=in.readLine();
        bookISBN=boi;
        editor=boe;
        //书的出版日期的输入,并判断输入格式是否正确
        boolean b=true;
        L3:
            while(b)
            {   System.out.print("请输入书的出版日期(yyyy-mm-dd):");
                bpds=in.readLine();
                try
                {   publishDate=bpddf.parse(bpds);
                    b=false;
                } catch (Exception e)
                {   System.out.print("输入日期不正确!请重试!\n");
                    continue L3;
                }
            }
    }catch (Exception e)
    {   System.out.print(e);
    }
}

//输出书的相关信息
public void getbookInfo()
```

```java
        {   //出版日期
            int year,month, day;
            year=publishDate.getYear()+1900;
                                    //函数输出值是与1900比较后的值,故要加1900
            month=publishDate.getMonth()+1;    //函数输出值为0~11,所以要加1
            day=publishDate.getDate();
            System.out.println();
            System.out.println("您输入了如下信息:\n"
                +"书的名称是:"+mediaName
                +"\n 书的价格是:"+price
                +"\n 书的作者是:"+artist
                +"\n 书的ISBN是:"+bookISBN
                +"\n 书的出版社是:"+press
                +"\n 书的编辑是:"+editor
                +"\n 书的出版日期是:"+year
                +"年"+month
                +"月"+day+"日");
            System.out.println();
        }
} //end of class MyBook

//CD子类
class MyCD extends MyMedia
{   String cdISRC;                          //CD的ISRC
    String cdPublisher;                     //CD的发行商

    //构造方法
    MyCD(String cn, float cp, String cpr, String ca, String ci, String cpl)
    {   mediaName=cn;
        price=cp;
        press=cpr;
        artist=ca;
        cdISRC=ci;
        cdPublisher=cpl;
    }

    //CD相关信息的输入成员函数
    public void cdOtherInfo ()
    {   try
        {   String cdi=new String();
            String cdp=new String();
            InputStreamReader ir=new InputStreamReader(System.in);
            BufferedReader in=new BufferedReader(ir);
            System.out.print("请输入CD的ISRC:");
```

```java
            cdi=in.readLine();
            System.out.print("请输入 CD 的发行商:");
            cdp=in.readLine();
            cdISRC=cdi;
            cdPublisher=cdp;
        }catch (Exception e)
        {   System.out.print(e);
        }
    }

    //输出 CD 的相关信息
    public void getcdInfo()
    {   System.out.println();
        System.out.println("您输入了如下信息:\n"
            +"CD 的名称是:"+mediaName
            +"\nCD 的价格是:"+price
            +"\nCD 的出版社是:"+press
            +"\n 出唱片的艺人是:"+artist
            +"\nCD 的 ISRC 是:"+cdISRC
            +"\nCD 的发行商是:"+cdPublisher);
        System.out.println();
    }
} //end of class MyCD

//磁带子类
class MyTape extends MyMedia
{   String tapeISRC;                            //磁带的 ISRC

    //构造方法
    MyTape(String tn, float tp, String ta, String ti, String tpr)
    {   mediaName=tn;
        price=tp;
        artist=ta;
        tapeISRC=ti;
        press=tpr;
    }

    //磁带相关信息的输入成员函数
    public void tapeOtherInfo ()
    {   try
        {   String tai=new String();
            InputStreamReader ir=new InputStreamReader(System.in);
            BufferedReader in=new BufferedReader(ir);
            System.out.print("请输入磁带的 ISRC:");
```

```java
                tai=in.readLine();
                tapeISRC=tai;
            }catch (Exception e)
            {   System.out.print(e);
            }
        }

        //输出磁带的相关信息
        public void gettapeInfo()
        {   System.out.println();
            System.out.println("您输入了如下信息:"
                +"\n 磁带的名称是:"+mediaName
                +"\n 磁带的价格是:"+price
                +"\n 出磁带的艺人是:"+artist
                +"\n 磁带的 ISRC 是:"+tapeISRC
                +"\n 磁带的发行商是:"+press );
            System.out.println();
        }
} //end of class MyTape

public class OperateMediaMain
{   public final static int MAX_ARRAY=200;
    public final static String MENU_TEXT[]=new String[]
        {   "请选择你要进行的操作(请输入 1~4 中的任一数字):\n"
            +"1:输入;\n"
            +"2:删除;\n"
            +"3:查找;\n"
            +"4:退出系统;\n",

            "请选择你要进行的操作(请输入 1~4 中的任一数字):\n"
            +"1:输入书的信息;\n"
            +"2:输入 CD 的信息;\n"
            +"3:输入磁带的信息;\n"
            +"4:返回上级目录;\n",

            "请选择你要进行的操作(请输入 1~4 中的任一数字):\n"
            +"1:删除书的信息;\n"
            +"2:删除 CD 的信息;\n"
            +"3:删除磁带的信息;\n"
            +"4:返回上级目录;\n",

            "请选择你要进行的操作(请输入 1~4 中的任一数字):\n"
            +"1:查找书的信息;\n"
            +"2:查找 CD 的信息;\n"
```

```java
        +"3:查找磁带的信息;\n"
        +"4:返回上级目录;\n",
    };

    //输入待删除的序列
    public int deleteNum()
    {   boolean b=true;
        InputStreamReader ir;
        BufferedReader in;
        String s=new String();
        int choice2=-1;

        try
        {   ir=new InputStreamReader(System.in);
            in=new BufferedReader(ir);
            while(b)
            {   //System.out.print("输入删除的序号:");
                System.out.flush();
                s=in.readLine();
                try
                {   choice2=Integer.parseInt(s);
                    if(choice2<0||choice2>=MAX_ARRAY)
                    {   System.out.println("输入错误!请重试:");
                        continue;
                    }
                    b=false;
                }catch(NumberFormatException e)
                {   System.out.println("非数字!请重试:");
                    continue;
                }
            }
        } catch (IOException e)
        {   System.out.println(e);
        }
        return choice2;
    }

    public static void main(String[] args)
    {   //定义3个一维数组,分别存放书、CD和磁带的信息
        MyBook BookInfo[]=new MyBook[MAX_ARRAY];
        MyCD CDInfo[]=new MyCD[MAX_ARRAY];
        MyTape TapeInfo[]=new MyTape[MAX_ARRAY];

        int BookKey[]=new int[MAX_ARRAY];
```

```java
int CDKey[]=new int[MAX_ARRAY];
int TapeKey[]=new int[MAX_ARRAY];
int firstBookkey=0;
int firstCDkey=0;
int firstTapekey=0;
int Bookcount=0;
int CDcount=0;
int Tapecount=0;

OperateMediaMain omm=new OperateMediaMain();

int choice=0;
boolean continu_e=true;
InputStreamReader ir;
BufferedReader in;
String s=new String();
int menuindex=0;
//初始化
for(int i=0; i<MAX_ARRAY; i++)
{    BookInfo[i]=null;
     CDInfo[i]=null;
     TapeInfo[i]=null;

     BookKey[i]=-1;
     CDKey[i]=-1;
     TapeKey[i]=-1;
}
firstBookkey=-1;
firstCDkey=-1;
firstTapekey=-1;
while(continu_e)
{    System.out.println(MENU_TEXT[menuindex]);
     boolean b=true;
     try
     {    ir=new InputStreamReader(System.in);
          in=new BufferedReader(ir);
          while(b)
          {    s=in.readLine();
               try
               {    choice=Integer.parseInt(s);
                    if(choice<1||choice>4)
                    {    System.out.println("输入错误!请重试:");
                         continue;
                    }
```

```
                b=false;
            }catch(NumberFormatException e)
            {   System.out.println("非数字!请重试:");
                continue;
            }
        }//end of while(b)
} catch (IOException e)
{   System.out.println(e);
}//end of try
switch(menuindex)
{   case 0://主菜单:
        menuindex=choice;
        if(menuindex==4)
            continu_e=false;
        break;
    case 1://插入
        int i;
        switch (choice)
        {   case 1:
                MyBook mb=new MyBook(" ", 0.0f, " ", " ", " ", " ");
                System.out.print("请输入书的名字:");
                mb.mediaNameInput();
                System.out.print("请输入书的价格:");
                mb.mediaPriceInput();
                System.out.print("请输入书的出版社:");
                mb.mediaPressInput();
                System.out.print("请输入书的作者:");
                mb.artistInput();
                //输入其他信息
                mb.bookOtherInfo();
                //输出用户输入的内容
                mb.getbookInfo();
                for(i=0;i<MAX_ARRAY;i++)
                    if(BookInfo[i]==null)
                        break;
                if(i==MAX_ARRAY)
                {   System.out.println("记录已满");
                    break;
                }
                BookInfo[i]=mb;
                //sort..
                if(Bookcount==0)
                {   firstBookkey=Bookcount;
                    BookKey [Bookcount]=-1;
```

```
                    }
                else
                {   int curi=firstBookkey;
                    int lasti=firstBookkey;
                    while (curi!=-1)
                    {   MyBook bk=BookInfo[curi];
                        if (bk.price<mb.price)
                        {   BookKey[i]=curi;
                            if (curi==lasti)
                            {   //list head
                                firstBookkey=i;
                            }
                            else
                            {   BookKey[lasti]=i;
                            }
                            break;
                        }//end of if (bk.price<mb.price)
                        lasti=curi;
                        curi=BookKey[curi];
                    }//end of while (curi!=-1)
                }//end of if(Bookcount==0)
            Bookcount++;
            break;
        case 2:
            MyCD mc=new MyCD(" ", 0.0f, " ", " ", " ", " ");
            System.out.print("请输入CD的名字:");
            mc.mediaNameInput();
            System.out.print("请输入CD的价格:");
            mc.mediaPriceInput();
            System.out.print("请输入CD的出版社:");
            mc.mediaPressInput();
            System.out.print("请输入出CD的艺人:");
            mc.artistInput();
            //输入其他信息
            mc.cdOtherInfo();
            //输出用户输入的内容
            mc.getcdInfo();
            for(i=0;i<MAX_ARRAY;i++)
                if(CDInfo[i]==null)
                    break;
            if(i==MAX_ARRAY)
            {   System.out.println("记录已满");
                break;
            }
```

```
            CDInfo[i]=mc;
            //sort..
            if(CDcount==0)
            {   firstCDkey=CDcount;
                CDKey [CDcount]=-1;
            }
            else
            {   int curi=firstCDkey;
                int lasti=firstCDkey;
                while (curi!=-1)
                {   MyCD cd=CDInfo[curi];
                    if (cd.price<mc.price)
                    {   CDKey[i]=curi;
                        if (curi==lasti)
                        { //list head
                            firstCDkey=i;
                        }
                        else
                        {   CDKey[lasti]=i;
                        }
                        break;
                    }//end of if (cd.price<mc.price)
                    lasti=curi;
                    curi=CDKey[curi];
                } //end of while (curi!=-1)
            }//end of if(CDcount==0)
            CDcount++;
            break;
        case 3:
            MyTape mt=new MyTape(" ", 0.0f, " ", " ", " ");
            System.out.print("请输入磁带的名字:");
            mt.mediaNameInput();
            System.out.print("请输入磁带的价格:");
            mt.mediaPriceInput();
            System.out.print("请输入磁带的发行商:");
            mt.mediaPressInput();
            System.out.print("请输入出磁带的艺人:");
            mt.artistInput();
            //输入其他信息
            mt.tapeOtherInfo();
            //输出用户输入的内容
            mt.gettapeInfo();
            for(i=0;i<MAX_ARRAY;i++)
                if(TapeInfo[i]==null)
```

```
                        break;
            if(i==MAX_ARRAY)
            {    System.out.println("记录已满");
                 break;
            }
            TapeInfo[i]=mt;
            //sort..
            if(Tapecount==0)
            {    firstTapekey=Tapecount;
                 TapeKey [Tapecount]=-1;
            }
            else
            {    int curi=firstTapekey;
                 int lasti=firstTapekey;
                 while (curi!=-1)
                 {   MyTape tk=TapeInfo[curi];
                     if (tk.price<mt.price)
                     {   TapeKey[i]=curi;
                         if (curi==lasti)
                         { //list head
                              firstTapekey=i;
                         }
                         else
                         {   TapeKey[lasti]=i;
                         }
                         break;
                     }//end of if (tk.price<mt.price)
                     lasti=curi;
                     curi=TapeKey[curi];
                 }//end of while (curi!=-1)
            }//end of if(Tapecount==0)
            Tapecount++;
            break;
        default:
            menuindex=0;
    } //end of switch (choice)
    break; //end of case 1://插入
case 2://删除
    int curi=firstBookkey;
    switch (choice)
    {   case 1:
            i=0;
            while(curi!=-1)
            {   i++;                            //输出书信息
```

```
            MyBook bk=BookInfo[curi];
            System.out.println("index : "+i
                +",at array index:"+curi);
            bk.getbookInfo();
            curi=BookKey[curi];
        } //end of while(curi!=-1)
        break;
    case 2:
        curi=firstCDkey;
        i=0;
        while(curi!=-1)
        {   i++;                        //输出书的信息
            MyCD bk=CDInfo[curi];
            System.out.println("index : "+i
                +",at array index:"+curi);
            bk.getcdInfo();
            curi=CDKey[curi];
        } //end of while(curi!=-1)
        break;
    case 3:
        curi=firstTapekey;
        i=0;
        while(curi!=-1)
        {   i++;                        //输出书的信息
            MyTape bk=TapeInfo[curi];
            System.out.println("index : "+i
                +",at array index:"+curi);
            bk.gettapeInfo();
            curi=TapeKey[curi];
        } //end of while(curi!=-1)
        //Tapecount++;
        break;
    default:
        menuindex=0;
        break;
} //end of switch (choice)
if(menuindex==0)
    break;
System.out.println("输入删除的序号:");
//函数实现:
int choice2=omm.deleteNum();
int lasti=firstBookkey;
i=0;
switch (choice)
```

```
        { case 1:
              lasti=curi=firstBookkey;
              while (curi!=-1)
              {   i++;
                  if(i==choice2)
                  {   if(curi==lasti)
                      {   //head
                          firstBookkey=BookKey[curi];
                      }
                      else
                      {   BookKey[lasti]=BookKey[curi];
                      }
                      BookInfo[curi]=null;
                      Bookcount--;
                      System.out.println("删除成功!");
                      break;
                  } //end of if(i==choice2)
                  lasti=curi;
                  curi=BookKey[curi];
              } //end of while (curi!=-1)
              if(curi==-1)
                  System.out.println("没找到记录,删除失败!");
              break;
          case 2:
              lasti=curi=firstCDkey;
              while (curi!=-1)
              {   i++;
                  if(i==choice2)
                  {   if(curi==lasti)
                      {   //head
                          firstCDkey=CDKey[curi];
                      }
                      else
                      {   CDKey[lasti]=CDKey[curi];
                      }
                      CDInfo[curi]=null;
                      CDcount--;
                      System.out.println("删除成功!");
                      break;
                  } //end of if(i==choice2)
                  lasti=curi;
                  curi=CDKey[curi];
              }//end of while (curi!=-1)
              if(curi==-1)
```

```
                System.out.println("没找到记录,删除失败!");
            break;
        case 3:
            lasti=curi=firstTapekey;
            while (curi!=-1)
            {   i++;
                if(i==choice2)
                {   if(curi==lasti)
                    {   //head
                        firstTapekey=TapeKey[curi];
                    }
                    else
                    {   TapeKey[lasti]=TapeKey[curi];
                    }
                    TapeInfo[curi]=null;
                    Tapecount--;
                    System.out.println("删除成功!");
                    break;
                }//end of if(i==choice2)
                lasti=curi;
                curi=TapeKey[curi];
            }//end of while (curi!=-1)
            if(curi==-1)
                System.out.println("没找到记录,删除失败 1!");
            break;
        default:
            menuindex=0;
            break;
    } //end of switch (choice)
    break;
case 3://查找
    b=true;
    if(choice==4)
    {   menuindex=0;
        break;
    }
    int choice2=-1;
    try
    {   ir=new InputStreamReader(System.in);
        in=new BufferedReader(ir);
        while(b)
        {   System.out.print("输入查找的最低价钱:");
            System.out.flush();
            s=in.readLine();
```

```java
            try
            {   choice2=Integer.parseInt(s);
                if(choice2<0)
                {   System.out.println("输入错误!请重试:");
                    continue;
                }
                b=false;
            }catch(NumberFormatException e)
            {   System.out.println("非数字!请重试:");
                continue;
            }
        } //end of while(b)
    } catch (IOException e)
    {   System.out.println(e);
    }
    b=true;
    int choice3=-1;
    try
    {   ir=new InputStreamReader(System.in);
        in=new BufferedReader(ir);
        while(b)
        {   System.out.print("输入查找的最高价钱:");
            System.out.flush();
            s=in.readLine();
            try
            {   choice3=Integer.parseInt(s);
                if(choice3<0)
                {   System.out.println("输入错误!请重试:");
                    continue;
                } //end of if(choice3<0)
                b=false;
            }catch(NumberFormatException e)
            {   System.out.println("非数字!请重试:");
                continue;
            }
        } //end of while(b)
    } catch (IOException e)
    {   System.out.println(e);
    }
    int curi=firstBookkey;
    int lasti=firstBookkey;
    switch (choice)
    {   case 1:
            curi=firstBookkey;
```

```
            lasti=firstBookkey;
            while (curi!=-1)
            {   MyBook bk=BookInfo[curi];
                if (bk.price >=choice2&& bk.price<=choice3 )
                    bk.getbookInfo();
                lasti=curi;
                curi=BookKey[curi];
            }
            break;
        case 2:
            curi=firstCDkey;
            lasti=firstCDkey;
            while (curi!=-1)
            {   MyCD bk=CDInfo[curi];
                if (bk.price >=choice2&& bk.price<=choice3 )
                    bk.getcdInfo();
                lasti=curi;
                curi=CDKey[curi];
            }
            break;
        case 3:
            curi=firstTapekey;
            lasti=firstTapekey;
            while (curi!=-1)
            {   MyTape bk=TapeInfo[curi];
                if (bk.price >=choice2&& bk.price<=choice3 )
                    bk.gettapeInfo();
                lasti=curi;
                curi=TapeKey[curi];
            } //end of while (curi!=-1)
            break;
        default:
            menuindex=0;
        } //end of switch (choice)
        break;//end of case 3://查找
    } //end of switch(menuindex)
  }//end of  while(continu_e)
   System.out.println("再见!\n");
}//end of main
boolean Insert(int choice)
{   boolean continu_e=true;
    return continu_e;
}
boolean Delete(int choice)
```

```
        {   return true;
        };
        boolean Search(int choice)
        {   return true;
        }
} //end of class OperateMediaMain
```

本例还使用向量作为存储结构实现了相关的功能,着重于对 Vector 类的练习。

与前例类似,设计了媒体类 MyMedia,包括的属性有:媒体名称 mediaName、价格 price、出版社(或发行商)press、作者(或艺人)artist;并定义了各自的赋值方法。从 MyMedia 派生各个子类,其中 MyBook 是书的子类,包括属性:ISBN、编辑;MyCD 是 CD 的子类,包括属性:ISRC 和发行商;MyTape 是磁带的子类,包括属性:ISRC。本例还实现了 print()方法,可以输出各类的属性,实现基本的打印功能。

同样地,设计了类 MyMediaManager,可以对媒体实例进行访问。分别定义了 3 个向量来对书、CD 及磁带的信息进行存储,也实现了插入、删除和查找功能。其中插入方法 addMedia(Vector medias, MyMedia tmpmedia)实现了在向量的末端插入一条新记录的功能,删除方法 deleteMedia(Vector medias)实现的是对某种媒体的删除,查找方法 findMedia(Vector medias, String mname)将按照媒体共有的属性即媒体名称来实现相应的查找。这里假定每种媒体中的各个名称是唯一的,当然也可以实现按照其他属性的查找,这些功能的实现留给读者完成。

使用向量作为存储结构的程序代码实现如下:

```
import java.*;
import java.io.*;
import java.util.Vector;

class MyMedia                                    //父类:Medida
{   String mediaName;                            //名称
    float price;                                 //价格
    String press;                                //出版社
    String artist;                               //艺术家(作者)

    MyMedia(String name,float price,String press,String artist)
    {   this.mediaName=name;
        this.price=price;
        this.press=press;
        this.artist=artist;
    }

    MyMedia(){ }

    void setName(String mname)
    {   this.mediaName=mname;
```

```java
    }
    void setPrice(float price)
    {    this.price=price;
    }
    void setPress(String press)
    {    this.press=press;
    }
    void setArtist(String artist)
    {    this.artist=artist;
    }

    void print()                                        //输出媒体
    {    System.out.print("Name:"+this.mediaName+"\t"+"Price:"+
            this.price+"\n"+"Press:"+this.press+"\t"+"Artist:"+this.artist+"\t");
    }
}; //end of class MyMedia

class MyCD extends MyMedia                              //CD 子类
{    String cdISRC;                                     //CD ISRC
    String cdPublisher;                                 //出版社
    MyCD(){};
    MyCD(String isrc, String name, float price, String press,
        String artist, String publisher)
    {    super(name, price, press, artist);
        this.cdISRC=isrc;
        this.cdPublisher=publisher;
    }
    void setCDISRC(String isrc)
    {    this.cdISRC=isrc;
    }
    void setCDPublisher(String publisher)
    {    this.cdPublisher=publisher;
    }

    void print()                                        //输出 CD 信息
    {    System.out.print("ISRC:"+this.cdISRC+"\t");
        super.print();                                  //调用父类的方法
        System.out.print("Publisher:"+this.cdPublisher+"\n");
    }
}; //end of class MyCD

class MyBook extends MyMedia                            //书子类
{    String editor;                                     //责任编辑
    String bookISBN;                                    //书的 ISBN
```

```java
        void setEditor(String editor)
        {   this.editor=editor;
        }
        void setBookISBN(String bookISBN)
        {   this.bookISBN=bookISBN;
        }
        MyBook(){};
        MyBook(String isbn, String name, float price, tring press, String artist,
        String editor)
        {   super(name, price, press, artist);
            this.bookISBN=isbn;
            this.editor=editor;
        }

        void print()                                        //输出书的信息
        {   System.out.print("ISBN:"+this.bookISBN+"\t");
            super.print();                                  //调用父类的方法
            System.out.print("editor:"+this.editor+"\n");
        }
};  //end of class MyBook

class MyTape extends MyMedia                                //磁带子类
{   String tapeISRC;                                        //磁带的 ISRC

        void setTapeISRC(String tapeISRC)
        {   this.tapeISRC=tapeISRC;
        }

        MyTape(){};
        MyTape(String isrc, String name, float price, String press, String artist)
        {   super(name, price, press, artist);
            tapeISRC=isrc;
        }

        void print()                                        //输出磁带的信息
        {   System.out.print("ISRC:"+this.tapeISRC+"\t");
            super.print();                                  //调用父类的方法
            System.out.println("");
        }
};  //end of class MyTape

public class MyMediaManager extends Vector                  //媒体库管理类
{   MyMediaManager(){ }
```

```
int addMedia(Vector medias, MyMedia tmpmedia)   //向媒体库添加媒体
{   medias.addElement(tmpmedia);
    return 1;
}

int deleteMedia(Vector medias)                  //删除媒体库中的媒体
{   if(medias.isEmpty()==false)
    {   int i =medias.size();
        medias.removeElementAt(i-1);
        return 1;
    }
    else
    {   return 0;
    }
}

MyMedia findMedia(Vector medias, String mname) //从媒体库中查找媒体,按名查找
{   MyMedia tmpmedia=new MyMedia();
    tmpmedia=null;
    int i=0;
    for(i=0;i<medias.size();i++)
    {   MyMedia getmedia= (MyMedia)medias.elementAt(i);
        if(((getmedia.mediaName).trim()).compareTo(mname.trim())==0)
        {   tmpmedia= (MyMedia)medias.elementAt(i);
        }
    }
    return tmpmedia;
}

public void printMedia(Vector medias)           //输出媒体库信息
{   MyMedia tmpcd;
    int length=medias.size();
    if(length>0)
    {   System.out.println("Number of vector elements is "+length+" and
        they are:");
        for (int i=0; i<length; i++)
        {   tmpcd= (MyMedia)medias.elementAt(i);
            tmpcd.print();
        }
    }
    else
    {   System.out.println("There is no one media ,please add.");
    }
```

```java
        }

        String inputstr()                                   //字符串输入函数
        {   String pm="";

            try
            {   InputStreamReader reader=new InputStreamReader(System.in);
                BufferedReader input=new BufferedReader(reader);
                pm =input.readLine();
            }catch (IOException e)
            {   System.out.println("IO Exception occur...");
            }
            return pm;
        }

        public static void main(String[] args)
        {   MyMediaManager cm=new MyMediaManager();
            Vector books=new Vector(50,1);                  //定义书向量,即书库
            Vector cds=new Vector(50,1);                    //定义CD向量,即CD库
            Vector tapes=new Vector(50,1);                  //定义磁带向量,即磁带库

            int count1=1;                                   //outer count
            int count=1;                                    //inner count
            //设置媒体选择菜单
            while(count1!=0)                                //outer loop
            {   System.out.print("----------------------------------\n ");
                System.out.print("which of following do you want to operate:(1,2,3,0)\n ");
                System.out.print("1.Book.\n ");
                System.out.print("2.CD.\n ");
                System.out.print("3.Tape.\n ");
                System.out.print("0.exit.\n ");
                try
                {   count1=Integer.parseInt(cm.inputstr());
                }catch(NumberFormatException ne)
                {   System.out.println("invalid data format ...");
                }
                switch (count1)
                {   case 1:                                 //对书籍进行操作
                        count=1;
                        while(count!=0)
                        {   System.out.print("------------------------------------------\n ");
                            System.out.print("which of following do
                                you want to do:(1,2,3,4,0)\n ");
```

```java
System.out.print("1.add book.\n ");
System.out.print("2.find book.\n ");
System.out.print("3.delete book.\n ");
System.out.print("4.print all books.\n ");
System.out.print("0.back.\n ");
try
{   count=Integer.parseInt(cm.inputstr());
}catch(NumberFormatException ne)
{   System.out.println("invalid data format ...");
}
switch (count)
{   case 1:                        //添加书
        //创建新的 MyBook 对象
        MyBook book1=new MyBook();
        //下面输入书的相关信息
        System.out.print("please input book ID : ");
        book1.setBookISBN(cm.inputstr());
        System.out.print("please input book Name : ");
        String mName="";
        boolean b=true;
        L1:    while(b)
            {   mName=cm.inputstr() ;
                if(mName.matches(""))
                    {   System.out.println("Name couldn't
                        be null ,retry:");
                        continue L1;
                    }
                b=false;
            }
        book1.setName(mName);

        System.out.print("please input book Price : ");
        float price=0;
        b=true;
        try
        {   while(b)
            {   price=Float.parseFloat(cm.inputstr());
                if(price <=0)
                {   System.out.print("Price should
                    be lager than 0,retry:");
                    continue ;
                }
                b=false;
            }
```

```java
            book1.setPrice(price);
        }catch(NumberFormatException ne)
        {   System.out.println("please input
               valid number!");
        }

        System.out.print("please input book Press : ");
        book1.setPress(cm.inputstr());
        System.out.print("please input book Author : ");
        book1.setArtist(cm.inputstr());
        System.out.print("please input book Editor: ");
        book1.setEditor(cm.inputstr());

        //将该书加入到书库中
        int i=cm.addMedia(books,book1);
        if(i>0)            //判断操作是否成功
        {   System.out.print("insert successful!\n ");
        }
        else
        {   System.out.print("insert failed!\n");
        }
        break;
    case 2:                //在书库中按书名来查找书
        System.out.println("please input Name of
            book you want to find: ");
        String name=cm.inputstr();
        //调用查找方法
        MyBook fbook= (MyBook)cm.findMedia(books,name);
        if (fbook==null)    //判断查找是否成功
        {   System.out.println("There is no book you
                want. ");
        }
        else
        {
            System.out.println("info of book you
                want to find is: ");
            fbook.print();
        }
        break;
    case 3:                //删除书
        if(cm.deleteMedia(books)>0)
        {   System.out.print("delete successful!\n ");
        }
        else
```

```
                    {  System.out.print("delete failed!\n ");
                    }
                    break;
                case 4:                          //输出书库的信息
                    cm.printMedia(books);
                    break;
                case 0:
                    break;
                default:
                    break;
            }//end of switch (count)
        }//end of while(count!=0)
        break;

    case 2:                                      //对CD进行管理
        count=1;
        while(count!=0)
        {   System.out.print("---------------------------
                    -----------\n ");
            System.out.print("which of following do you
                    want to do:(1,2,3,4,0)\n ");
            System.out.print("1.add cd.\n ");
            System.out.print("2.find cd.\n ");
            System.out.print("3.delete cd.\n ");
            System.out.print("4.print all cds.\n ");
            System.out.print("0.back.\n ");
            try
            {   count=Integer.parseInt(cm.inputstr());
            }catch(NumberFormatException ne)
            {   System.out.println("invalid data format ...");
            }
            switch (count)
            {   case 1:                          //向CD库中添加新的CD
                    MyCD cd1=new MyCD();        //创建新的MyCD对象
                    //下面输入cd1的相关信息
                    System.out.print("please input cd ID : ");
                    cd1.setCDISRC(cm.inputstr());

                    System.out.print("please input cd Name : ");
                    String mName="";
                    boolean b=true;
                    L1:
                        while(b)
```

```java
            {   mName =cm.inputstr() ;
                if(mName.matches(""))
                {   System.out.println("Name
                    couldn't be null ,retry:");
                    continue L1;
                }
                b=false;
            }
        cd1.setName(mName);

        System.out.print("please input cd Price : ");
        float price=0;
        b=true;
        try
        {   while(b)
            {   price=Float.parseFloat(cm.inputstr());
                if(price <=0)
                {   System.out.print("Price should be
                    lager than 0,retry:");
                    continue ;
                }
                b=false;
            }//end of while(b)
            cd1.setPrice(price);
        }catch(NumberFormatException ne)
        {   System.out.println("please input valid
                number!");
        }

        System.out.print("please input cd Press : ");
        cd1.setPress(cm.inputstr());
        System.out.print("please input cd Artist : ");
        cd1.setArtist(cm.inputstr());
        System.out.print("please input cd Publisher: ");
        cd1.setCDPublisher(cm.inputstr());

        int i=cm.addMedia(cds,cd1);   //将 cd1 放入 CD 库

        if(i>0)                                //判断操作是否成功
        {   System.out.print("insert successful!\n ");
        }
        else
        {   System.out.print("insert failed!\n");
```

```
            }
            break;
        case 2:                      //以名称来查找CD
            System.out.println("please input Name of cd you
                want to find: ");
            String name=cm.inputstr();
            MyCD fcd=(MyCD)cm.findMedia(cds,name); //find cd
            if (fcd==null)           //判断查找是否成功
            {   System.out.println("There is no cd
                    you want. ");
            }
            else
            {   System.out.println("info of cd you want to
                    find is: ");
                fcd.print();
            }
            break;
        case 3:                      //删除CD
            if(cm.deleteMedia(cds)>0)
            {   System.out.print("delete successful!\n ");
            }
            else
            {   System.out.print("delete failed!\n ");
            }
            break;
        case 4:                      //输出CD库中的信息
            cm.printMedia(cds);
            break;
        case 0:
            break;
        default:
            break;
        }//end of switch(count)
    }//end of while(count!=0)
    break;

case 3:                              //对磁带进行操作
    count=1;
    while(count!=0)
    {   System.out.print("-----------------------------
            ------------------\n ");
        System.out.print("which of following do you want
            to do:(1,2,3,4,0)\n ");
```

```
System.out.print("1.add tape.\n ");
System.out.print("2.find tape.\n ");
System.out.print("3.delete tape.\n ");
System.out.print("4.print all tapes.\n ");
System.out.print("0.back.\n ");
try
{   count=Integer.parseInt(cm.inputstr());
}    catch(NumberFormatException ne)
{   System.out.println("invalid data format ...");
}
switch (count)
{   case 1:                        //添加磁带
            //创建一个新的 MyTape 对象
            MyTape tape1=new MyTape();
            //输入其相关信息
            System.out.print("please input tape ID : ");
            tape1.setTapeISRC(cm.inputstr());

            System.out.print("please input tape Name : ");
            String mName="";
            boolean b=true;
            L1:    while(b)
                {   mName =cm.inputstr() ;
                    if(mName.matches(""))
                    {   System.out.println("Name couldn't
                        be null ,retry:");
                        continue L1;
                    }
                    b=false;
                }
            tape1.setName(mName);

            System.out.print("please input tape Price : ");
            float price=0;
            b=true;
            try
            {   while(b)
                {   price=Float.parseFloat(cm.inputstr());
                    if(price <=0)
                    {   System.out.print("Price should be
                        lager than 0,retry:");
                        continue ;
                    }
                    b=false;
                }
            tape1.setPrice(price);
```

```java
        }catch(NumberFormatException ne)
        {   System.out.println("please input valid
                number!");
        }

        System.out.print("please input tape Press : ");
        tape1.setPress(cm.inputstr());
        System.out.print("please input tape Artist : ");
        tape1.setArtist(cm.inputstr());

        //将tape1加入到磁带库
        int i=cm.addMedia(tapes,tape1);
        if(i>0)                    //判断操作是否成功
        {   System.out.print("insert successful!\n ");
        }
        else
        {   System.out.print("insert failed!\n");
        }
        break;
    case 2:                        //以名字来查找磁带
        System.out.println("please input Name of tape
            you want to find: ");
        String name=cm.inputstr();
        //查找磁带
        MyTape ftape=(MyTape)cm.findMedia(tapes,name);
        if (ftape==null)       //判断查找是否成功
        {   System.out.println("There is no tape
                you want. ");
        }
        else
        {   System.out.println("info of tape you
                want to find is: ");
            ftape.print();
        }
        break;
    case 3:                        //删除磁带
        if(cm.deleteMedia(tapes)>0)
        {   System.out.print("delete successful!\n ");
        }
        else
        {   System.out.print("delete failed!\n ");
        }
        break;
    case 4:                        //输出磁带信息
        cm.printMedia(tapes);
        break;
```

```
                    case 0:
                        break;
                    default:
                        break;
                }//end of switch (count)
            }//end of while(count!=0)
            break;

        case 0:                                    //退出
            System.exit(0);
        default:
            break;
        }//end of outer switch (count1)
    }//end of outer while (count1!=0)
}//end of main
};
```

下面对向量实现进行测试,以磁带为例。

首先在主菜单中选择 3,进入磁带处理操作。选择 1.add tape,添加一盘磁带,选项执行的结果如图 5-2 所示。

图 5-2 插入一个新的磁带

这里,依次插入两条记录,选择 4.print all tape,查看执行结果,如图 5-3 所示。

之后,选择 2.find tape 来查询磁带,执行结果如图 5-4 所示。

最后,测试删除功能,选择 3.delete tape ,删除磁带,执行结果如图 5-5 所示。操作完毕,再次查看删除后的结果,如图 5-6 所示。

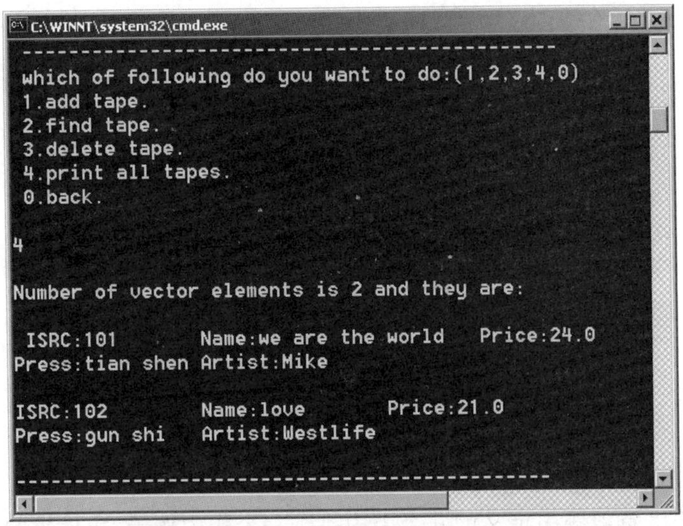

图 5-3 查看 tape 目前的记录

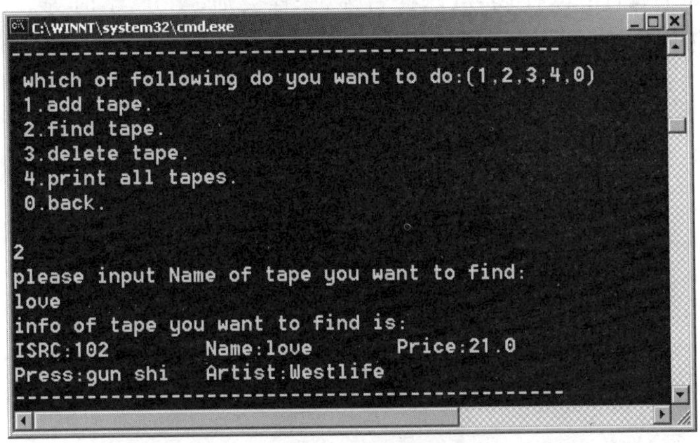

图 5-4 查询磁带

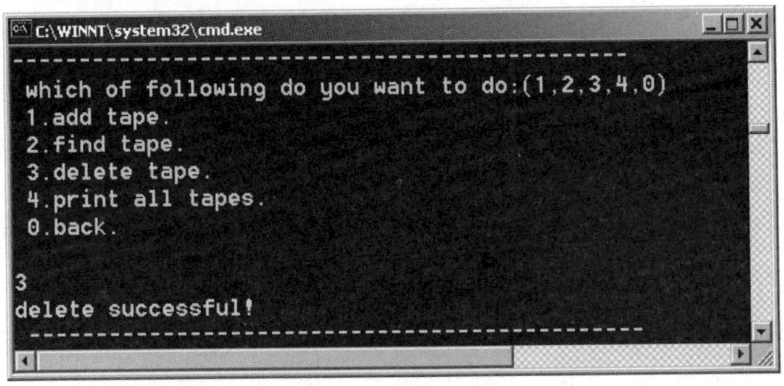

图 5-5 删除磁带

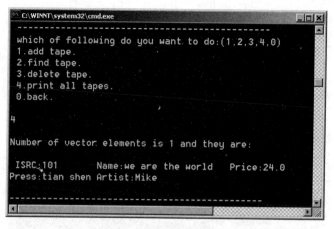

图 5-6　查看删除后的结果

5.19　从第 2 章习题中定义的几个类中挑选一个作为例子，完成类的定义。

解：这里选择 Course 类来完成类的定义。Course 类代表学校中的一门课程。一门课程一般包括如下属性：课程代号、课程名称、课程类别、学分、学时等。在类中我们定义了各个属性的访问方法。具体实现过程参见习题 2.11。

第6章 Java 语言中的异常

6.1 什么是异常？解释"抛出""捕获"的含义。

解：Java 把程序运行中可能遇到的错误分为两类，一类是非致命性的，通过某种修正后程序还能继续执行。这类错误称作异常，这是一类特殊的运行错误对象。比如打开一个文件时，发现文件不存在就属于这一类。这类错误并不会使程序瘫痪，只要得到适当处理，程序还会继续执行。比如，当发现文件不存在时，提示用户找到正确的文件名，程序还可以继续执行。另一类是致命性的，即程序遇到了非常严重的不正常状态，不能简单地恢复执行，这就是错误。

有些书上也把异常称为例外。为了能够及时有效地处理程序中的运行错误，Java 中引入了异常和异常类，以对这些运行错误进行统一处理。

程序运行过程中发生一个可识别的运行错误时，称程序产生了一个异常事件。对于这些可识别的异常，系统都有一个异常类与之对应，同时系统相应地会生成一个对应于该异常类的异常对象。该对象可能由正在运行的方法生成，也可能由 JVM 生成。异常对象中包含了必要的信息，包括所发生的异常事件的类型及异常发生时程序的运行状态。生成的异常对象传递给 Java 运行时系统，通过相应的机制来处理它，确保不会产生非正常中断等情况。异常分两类，一类是系统定义的，另一类是用户自定义的。

针对 Java 程序中引发的可识别错误，产生与该错误相对应的异常类对象的这一过程称为抛出。系统定义的所有运行异常都由系统自动抛出，用户程序自定义的异常由 throw 语句来抛出。

当抛出一个异常时，有三种方法可以处理抛出的异常：可以忽略它，这将导致程序中断；可以使用 try 语句在发生异常的地方处理它；或是 Java 运行时系统从生成对象的代码块开始，沿方法的调用栈逐层回溯，在调用层次中更高层的方法中捕获并处理它。寻找相应的处理代码，并把异常对象交给相应方法进行处理的这一过程称为捕获。简言之，专门处理异常的过程称为捕获。Java 程序中，catch 语句负责捕获异常。

6.2 Java 是如何处理异常的？

解：当发生异常时，要进行异常处理。Java 语言提供了异常处理机制，用于专门处理异常。一般地，当发生异常时，程序中断执行，并输出一条信息。Java 中，使用 try 语句括住可能抛出异常的代码段，用 catch 语句指明要捕获的异常及相应的处理代码。

try 语句与 catch 语句的语法格式如下：

```
try
{   //此处为抛出具体异常的代码
} catch (ExceptionType1 e)
{   //抛出 ExceptionType1 异常时要执行的代码
```

```
    } catch (ExceptionType2 e)
    {    //抛出 ExceptionType2 异常时要执行的代码
    …
    } catch (ExceptionTypek e)
    {    //抛出 ExceptionTypek 异常时要执行的代码
    }finally
    {    //必须执行的代码
    }
```

其中，ExceptionType1、ExceptionType2、…、ExceptionTypek 是产生的异常类型。根据发生异常所属的类，找到对应的 catch 语句，然后执行其后的语句序列，完成对异常的处理，恢复程序的执行。

6.3 catch 及 try 语句的作用是什么？语法格式如何？

解：对于可能抛出异常的代码段，要使用 try 语句括住，用 catch 语句指明要捕获的异常及相应的处理代码。

try 语句与 catch 语句的语法格式如下：

```
try
{    //此处为抛出具体异常的代码
} catch (ExceptionType1 e)
{    //抛出 ExceptionType1 异常时要执行的代码
} catch (ExceptionType2 e)
{    //抛出 ExceptionType2 异常时要执行的代码
…
} catch (ExceptionTypek e)
{    //抛出 ExceptionTypek 异常时要执行的代码
}finally
{    //必须执行的代码
}
```

其中，ExceptionType1、ExceptionType2、…、ExceptionTypek 是产生的异常类型。根据发生异常所属的类，找到对应的 catch 语句，然后执行其后的语句序列，完成对异常的处理，恢复程序的执行。

6.4 在什么情况下执行 try 语句中 finally 后面的代码段？在什么情况下不执行？试举例说明。

解：在执行 try 语句时，不论是否捕获到异常，都要执行 finally 后面的语句。try 后大括号（{ }）中的代码称为保护代码。如果在保护代码内执行了 System.exit()方法，程序将退出，此时不执行 finally 后面的语句，这是不执行 finally 后面语句的唯一一种可能。

下面的例子将从键盘读入一行信息，根据读入的内容来判定是否发生了异常。如果读入的是空串，则抛出 EmptyStringException()异常；如果读入的内容中含有数字，则抛出 IncludeNumberException()异常。程序中用到了 String 的 indexOf()函数，它返回参数所指定的字符在字符串中第一次出现的位置。如果这个位置大于等于 0，表明字符串中含有该字符。程序中使用循环来查找是否出现 0~9 这十个数字。

程序中使用了 finally 语句,输出一条信息：Exception example end。

程序代码实现如下：

```java
import java.io.InputStream;
import java.io.BufferedInputStream;
import java.io.*;

//定义的两个异常类
//当输入一个空串时的异常类
class EmptyStringException extends Exception {}
//当字符串中含有数字时的异常类
class IncludeNumberException extends Exception {}

public class ExceptionExample
{   public static void main(String [] args)
    {   BufferedReader in=new BufferedReader(
            new InputStreamReader(System.in));

        System.out.println("Please input : ");
        String str="";
        Try
        {   str=in.readLine();                          //从键盘读入一行信息
            //读入空串,则抛出 EmptyStringException()异常
            if(str.length()==0) throw new EmptyStringException();
            for(int i=0; i<10; i++)
                //依次判定读入的串中是否含有数字,有则抛出 IncludeNumberException()
                if( str.indexOf ( i+'0') >=0) throw new IncludeNumberException();
        } catch (EmptyStringException e)                //空串的处理
        {   System.out.println( "EmptyStringException" );
        } catch (IncludeNumberException e )             //含数字时的处理
        {   System.out.println( "IncludeNumberException" );
        } catch (IOException e )                        //其他输入异常的处理
        {   System.out.println( "IOException" );
        }finally                                        //永远执行的部分
        {   System.out.println( "Exception example end" );
        }
    }
}
```

上述程序的执行结果如图 6-1 所示。

修改保护代码段,增加一条退出语句,如下所示：

```java
try
{   str=in.readLine();
```

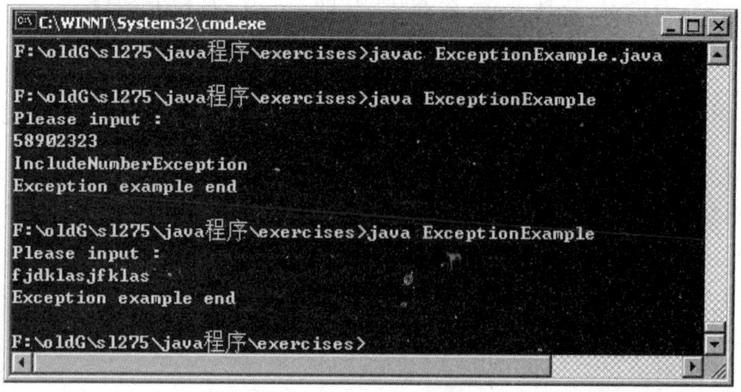

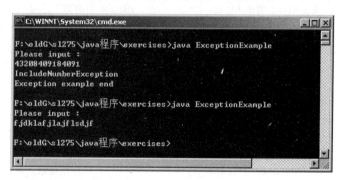

图 6-1 异常示例之一

```
        if(str.length()==0) throw new EmptyStringException();
        for(int i=0; i<10; i++)
            if(str.indexOf(i+'0') >=0) throw new IncludeNumberException();
        System.exit(0);
    }
```

当读入的内容不抛出异常时,将执行 System.exit(0)语句,从程序中退出,此时不会输出"Exception example end"信息。但当抛出异常时,程序转去处理异常,这样就会绕过 System.exit(0)语句。根据异常的类型执行不同的操作,并在最后输出"Exception example end"信息。

添加了 System.exit(0)语句后执行的结果如图 6-2 所示。

图 6-2 异常示例之二

6.5 你能说出 Java 中常见的几个异常吗?它们表示什么意思?在什么情况下引起这些异常?

解:Java 中定义了一些常见异常,这些是系统预定义的,它们的处理由系统自动执行。常见的几个异常有:

(1) ArithmeticException 算术异常

整数除法中,如果除数为 0,则发生该类异常。Java 虚拟机遇到这样的错误时会自动中止程序的执行流程,并新建一个 ArithmeticException 类的对象,即抛出一个算术运算异常。例如下面的程序将引发 ArithmeticException 异常:

```
public class TestArithmeticException
{   public static void main(String args[])
    {   int denominator=0, numerator=20, result=0;
        result=numerator /denominator;
        //除数为零,将引发 ArithmeticException 异常
        System.out.println( result );
    }
}
```

（2）NullPointerException 空指针异常

如果一个对象还没有实例化,那么访问该对象或调用它的方法时将导致 NullPointerException 异常。因此使用一个变量前,要先检查一下它是否为 null。

例如：

```
image im[]=new image [4];
System.out.println(im[0].toString());
```

第一行创建了有 4 个元素的数组 im,每个元素都是 image 类型的。系统为其进行初始化,每个元素中的值为 null,表明它还没有指向任何实例。第二行要访问 im[0],由于访问的是还没有进行实例化的空引用,因此导致 NullPointerException 异常。

（3）NegativeArraySizeException 数组元素个数为负异常

按常规,数组的元素个数应是一个大于等于 0 的整数,不应该是一个负数。创建数组时,如果元素个数是个负数,则会引发 NegativeArraySizeException 异常。

（4）ArrayIndexOutOfBoundsException 数组下标越界异常

Java 把数组看作是对象,并用 length 变量记录数组的大小。访问数组元素时,运行时环境根据 length 值自动检查下标的大小。如果数组下标越界,则将导致 ArrayIndexOutOfBoundsException 异常。

（5）SecurityException 安全异常

该类异常一般在浏览器内抛出。若 Applet 试图进行下述操作,则由 SecurityManager 类抛出此异常：

- 访问本地文件。
- 打开一个套接口,而不是返回到提供 Applet 的主机。
- 在运行时环境中运行另一个程序。

除此之外,常见异常还有：

- ArrayStoreException：程序试图存取数组中错误的数据类型。
- FileNotFoundException：试图存取一个并不存在的文件。
- IOException：通常的 I/O 错误。

6.6 请看下面的定义：

```
String s=null
(1) if ((s!=null) & (s.length()>0))
(2) if ((s!=null) && (s.length()>0))
```

(3) if ((s==null)|(s.length()==0))

(4) if ((s==null)||(s.length()==0))

在上面4个语句中,哪个能引发异常?是哪种类型的异常?

解:第一种情况,变量 s 的初值是 null,这表明 s 还没有实例化,那么不能调用它的方法。所以调用 s.length()将导致 NullPointerException 异常。编译含有该行的程序,将输出如下的出错信息:

```
Exception in thread "main" java.lang.NullPointerException
    at TestException.main(TestException.java:6)
```

第二种情况,条件中使用的是逻辑与 &&,它具有短路功能,即如果其左侧的条件不满足时,不再判定其右侧的条件,表达式返回假值。这也正是第二种情况的执行顺序。当(s!=null)不成立时,不再调用 s.length(),因此不会抛出异常。if 语句的条件为假,不执行其后的语句。

第三种情况,与第一种情况类似,条件中使用的是按位或"|",它不具有短路功能,因此运算符两侧的条件均需判别。调用 s.length()将导致 NullPointerException 异常。

第四种情况,与第二种情况类似,条件中使用的是逻辑或"||",它也具有短路功能,即如果其左侧的条件为真,不再判定其右侧的条件,表达式返回真值。本例中,(s==null)成立,所以 if 语句的条件为真,并且不再调用 s.length(),不会抛出异常。

6.7 查阅 API 文档,找出数组操作可能引起的异常。

解:数组操作可能引起的异常有:

(1) NegativeArraySizeException:按常规,数组的元素个数应是一个大于等于 0 的整数。创建数组时,如果元素个数是个负数,则会引发 NegativeArraySizeException 异常。

(2) IllegalArgumentException:若数组的起始下标大于终止下标,可能引发无效参数 IllegalArgumentException 异常。

(3) ArrayIndexOutOfBoundsException:若数组的起始下标小于 0,或者终止下标大于数组的长度,可能引发数组下标越界 ArrayIndexOutOfBoundsException 异常。

(4) ClassCastException:在进行数组比较或数组元素的比较时,如果数组中的元素相互没有可比性,或查询的关键字和数组的元素之间没有可比性,可能引发 ClassCastException 异常。

6.8 对程序 5-13(见主教材)实现的程序增加异常处理。

解:程序 5-13 的代码是关于事件处理的代码,对鼠标事件进行了监控。为此,对运行环境添加 HeadlessException 异常处理。当含有键盘键入、鼠标单击等操作的程序运行在不提供相应设备的环境下时,会抛出 HeadlessException 异常。

程序代码实现如下:

```
import java.awt.*;
import java.awt.event.*;
```

```java
public class TwoListenInner{
    private Frame f;
    private TextField tf;

    public static void main(String ars[])
    {   TwoListenInner that=new TwoListenInner();
        that.go();
    }
    public void go()
    {   try
        {   f=new Frame("Two listeners example");
            f.add("North", new Label("Click and drag the mouse"));
            tf=new TextField (30);
        }catch(HeadlessException e)
        {   System.out.println(e.getMessage());
        }
        f.add("South", tf);
        f.addMouseMotionListener(new MouseMotionHandler());
        f.addMouseListener(new MouseEventHandler());
        f.setSize(300, 200);
        f.setVisible(true);
    }
    //MouseMotionHandler 为一个内部类
    public class MouseMotionHandler extends MouseMotionAdapter{
        public void mouseDragged (MouseEvent e){
            String s="Mouse dragging: X="+e.getX() +"Y="+e.getY();
            tf.setText(s);
        }
    }
    //MouseEventHandler
    public class MouseEventHandler extends MouseAdapter{
        public void mouseEntered(MouseEvent e){
            String s="The mouse entered";
            tf.setText(s);
        }
        public void mouseExited (MouseEvent e){
            String s="The mouse has left the building";
            tf.setText(s);
        }
    }
}
```

第7章 Java 语言的高级特性

7.1 什么是泛型？为什么要使用泛型？

解：在设计类和接口时，需要说明相关的数据类型。Java 语言允许在类或接口的定义中，用一个占位符替代实际的类类型。这个技术称为泛型（Generic type 或者 generics）。

泛型是对 Java 语言的类型系统的一种扩展，以支持创建可以按类型进行参数化的类。可以把类型参数看作是使用参数化类型时指定的类型的一个占位符，就像方法的形式参数是运行时传递的值的占位符一样。通过使用泛型，可以定义一个类，其对象的数据类型由类的使用者在以后来确定。这样定义的类或接口，适用性更广。

使用泛型有以下目的：

(1) 类型安全。泛型的主要目标是提高 Java 程序的类型安全。通过知道使用泛型定义的变量的类型限制，编译器可以在多数情况下尽可能地验证类型假设。没有泛型，这些假设就只存在于程序员的头脑中。泛型的实现方式，使得所有的工作都在编译器中完成，编译器生成类似于没有泛型（和强制类型转换）时所写的代码，更能确保类型安全。

(2) 消除强制类型转换。泛型的一个附带好处是，消除源代码中的许多强制类型转换。这使得代码更加可读，并且减少了出错机会。

(3) 潜在的性能收益。泛型可能带来更高程度的优化。在泛型的初始实现中，编译器将强制类型转换插入到生成的字节码中。由于泛型的实现方式，JVM 所做的事情更少，几乎不需要更改类文件了。

7.2 如何借助于泛型定义一个线性表？

注：线性表是由相同类型的对象组成的一个线性结构。

解：可以参考如下示例：

```
import java.util.ArrayList;
import java.util.List;

public class ListDemo {
    public static void main(String[] args) {
        int numLength=10;
        int deleteNum=5;
        List<Integer>list=new ArrayList<Integer>();
        init(numLength,list);
        delete(deleteNum,list);
        print(list);
    }
```

```
        private static void print(List<Integer>list) {
            for(int i=0;i<list.size();i++){
                System.out.print(list.get(i)+"\t");
            }
        }

        private static void delete(int deleteNum,List<Integer>list) {
            for (int i=0;i<list.size();i++){
                if((int)list.get(i)==deleteNum){
                    list.remove(i);
                }
            }
        }

        private static void init(int numLength,List<Integer>list) {
            for(int i=1;i<=numLength;i++){
                list.add(i);
            }
        }
    }
```

7.3 像 String 或 Name 这样的类必须定义哪些方法,才能让 OrderedPair 的方法 toString 正常工作?

解:toString 方法。

主教材例 7.2 中给出了类 OrderedPair 的实现,其中,OrderedPair<T>实现了 Pairable<T>接口。类 OrderedPair 的方法 toString 如下:

```
public String toString()
{
    return "("+first+", "+second+")";
} //end toString
```

故在 String 或 Name 这样的类中,也必须定义 toString 方法。

7.4 考虑类 OrderedPair。假定没有使用泛型,而是忽略<T>,将私有域、方法参数及局部变量的数据类型声明为 Object 而不是 T。这些修改对类的使用有什么影响?

解:如果没有使用泛型,将私有域、方法参数及局部变量的数据类型声明为 Object 而不是 T,则使用不同的及不相关类型的对象组成对时,编译器不能提示错误信息,这样不能保证类型的正确使用。

例如,下列语句是合理的:

```
OrderedPair fruit=new OrderedPair("apple", "banana");
```

但如下的语句是不合理的：

```
Name joe=new Name("Joe", "Java");
String joePhone="(401) 555-1234";
OrderedPair joeEntry=new OrderedPair();
joeEntry.setPair(joe, joePhone);
```

如果不使用泛型，则对于这样的语句，编译器是检查不出错误的。

7.5 能使用类 OrderedPair，让两个不同及不相关的数据类型的对象配对吗？请解释原因。

解：不能让两个不同及不相关的数据类型的对象配对。

OrderedPair 类的构造方法的定义是这样的：

```
public OrderedPair(T firstItem, T secondItem)    //注：构造方法名后没有<T>
{
    first=firstItem;
    second=secondItem;
} //end constructor
```

其中，对配对的两个参数定义的是同一个泛型。

7.6 使用类 Name，写语句，将两名学生组成实验搭档。

解：可以使用如下的语句，将两个学生组成一个 OrderedPair 对象。

```
Name kristen=new Name("Kristen", "Doe");
Name luci=new Name("Luci", "Lei");
OrderedPair<Name>labPartners=new OrderedPair<>(kristen, luci);
```

7.7 定义泛型方法 swap，交换所给数组中两个指定位置的对象。

解：参考代码如下：

```
/** Interchanges two entries at given position within a given array. */
public static <T>void swap(T[] a, int i, int j)
{
    T temp=a[i];
    a[i]=a[j];
    a[j]=temp;
} // end swap
```

7.8 若集合中含有 4 个元素，则游标的可能位置有几个？分别位于什么位置？使用一个图来表示。

解：若集合中有 4 个元素，则游标的可能位置有 5 个，分别是第 1 个元素之前、最后一个元素之后及两个相邻元素之间，如图 7-1 所示。

7.9 假定有例 7-4 所示的 nameList，其中含有名字 Jamie、Joey 和 Rachel。下列 Java 语句会得到什么输出？

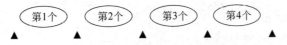

图 7-1　迭代器中可能的游标位置

```
Iterator<String>nameIterator=namelist.iterator();
nameIterator.next();
nameIterator.next();
nameIterator.remove();
System.out.println(nameIterator.hasNext());
System.out.println(nameIterator.next());
```

解：得到的输出是：

true
Rachel

前两个 next() 方法的调用，使得游标位于 Joey 和 Rachel 之间。调用 remove() 方法的结果是删除了 Joey。此时，迭代器游标还没有到达结尾处，故 hasNext() 方法的返回值是 true。再调用 next() 方法时，游标位于 Rachel 之后，返回的字符串是 Rachel，并输出。

7.10　如果 nameList 中至少含有 3 个字符串，且 nameIterator 如习题 7.9 中所定义的，写出 Java 语句，显示 nameList 的第 3 项。

解：语句序列如下：

```
Iterator<String>nameIterator=namelist.iterator();
nameIterator.next();
nameIterator.next();
System.out.println(nameIterator.next());
```

7.11　假定 nameList 和 nameIterator 如习题 7.9 和 7.10 所给出，写出语句，显示线性表中的偶数项。即显示第二项、第四项，等等。

解：语句序列如下：

```
Iterator<String>nameIterator=namelist.iterator();
if(nameIterator.hasNext()) nameIterator.next();         //跳过第一项
while(nameIterator.hasNext())
{    System.out.println(nameIterator.next());           //显示偶数项
    if(nameIterator.hasNext())
        nameIterator.next();                            //跳过奇数项
}
```

7.12　假定 nameList 和 nameIterator 如习题 7.9 和 7.10 所给出，写出语句，删除线性表中的所有项。

解：

```
Iterator<String>nameIterator=namelist.iterator();
while (nameIterator.hasNext())
```

```
    {
        nameIterator.next();
        nameIterator.remove();
} //end while
```

【拓展思考】

（1）假定，traverse 是如下定义的迭代器：

```
Iterator<String>traverse=namelist.iterator();
```

但 nameList 的内容未知。写 Java 语句，按反序显示 nameList 中的名字，从表尾开始。

解：

```
Iterator<String>traverse=namelist.iterator();
while (traverse.hasNext())
    traverse.next();                      //前进到表尾
while (traverse.hasPrevious())            //反向迭代
    System.out.println(traverse.previous());
```

（2）traverse 的定义如下：

```
Iterator<String>traverse=namelist.iterator();
```

namelist 中含有名字 Jen、Jim 和 Josh，此时迭代器位置位于 namelist 的头两项之间，写 Java 语句，用 Jon 替换 Josh。

解：

```
Iterator<String>traverse=namelist.iterator();
traverse.next();              //返回 Jim,游标位于 Jim 与 Josh 之间
traverse.next();              //返回 Josh,游标位于 Josh 之后
traverse.set("Jon");          //替换 Josh
```

（3）namelist 中含有名字 Jen、Ashley、Jim 和 Josh。如果迭代器位置位于 Ashley 和 Jim 之间，写 Java 语句，紧接在 Jim 的后面添加 Miguel。

解：

```
Iterator<String>traverse=namelist.iterator();
traverse.next();              //返回 Jim
traverse.add("Miguel");       //在 Jim 之后添加 Miguel
```

7.13 请解释什么是克隆、浅克隆及深克隆？

解： 在 Java 中，克隆是对象的拷贝。有些教材中，克隆也称为复制。

对象的克隆分为深克隆和浅克隆。浅克隆将对象中的数值类型的字段克隆到新的对象中，而对于对象中的引用型字段，则是将它的一个引用克隆到目标对象中。如果改变目标对象中引用型字段的值，也会反映在原始对象中，即原始对象中对应的字段也会发生变化。

深克隆是指源对象与克隆对象之间互相独立，在新对象中创建一个新的和原始对象

中对应字段相同（内容相同）的字段，其中任何一个对象的改动都不会对另外一个对象造成影响。深克隆不仅克隆对象的基本类，同时也克隆原对象中的对象。

7.14 参照 7.3 节中给出的 Student 类的定义，假定 x 是 Student 的一个实例，而 y 是它的克隆；即

```
3Student y=(Student)x.clone();
```

如果执行下列语句改变 x 的姓：

```
Name xName=x.getName();
xName.setLast("Smith");
```

那么 y 的姓会改变吗？请解释原因。

解：不能改变。

为 Student 类定义的 clone 方法如下：

```
public Object clone()
{
    Student theCopy=null;
    try
    {
        theCopy=(Student)super.clone();         //Object can throw an exception
    }
    catch (CloneNotSupportedException e)
    {
        throw new Error(e.toString());
    }
    theCopy.fullName=(Name)fullName.clone();
    return theCopy;
} //end clone
```

语句 theCopy.fullName＝(Name)fullName.clone()；表示这里进行的是深克隆，即对对象内含有的对象也进行了克隆。

克隆 y 有一个 Name 对象，它不同于 x 的 Name 对象，对 x 的修改不会影响到 y 的值，如图 7-2 所示。

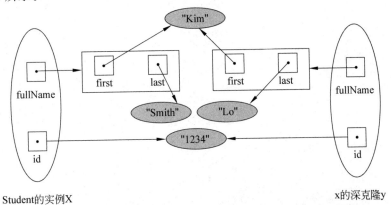

图 7-2　Student 和其深克隆的实例

7.15 在习题 7.14 的基础上，继续回答：如果没在 Student 的 clone 方法内克隆 fullName，修改 x 的姓也会改变 y 的姓吗？请解释原因。

解：会改变 y 的姓。没有这条语句，表示只对对象进行克隆，而对象内的引用并没有克隆，进行的是浅克隆，此时两个对象共享同一个 Name 对象，如图 7-3 所示。

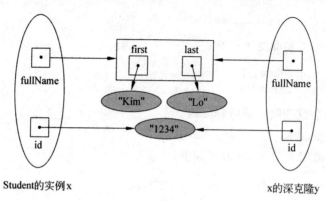

图 7-3 Student 和其浅克隆的实例

第8章 Java 的图形用户界面设计

8.1 编写程序创建并显示一个标题为"My Frame"、背景为红色的 Frame。

解：JFrame 是一个带有标题行和控制按钮（最小化、恢复/最大化、关闭）的独立窗口，创建应用程序时需要使用 JFrame。

JFrame 类的构造方法为：

```
JFrame frame=new JFrame(String);
```

其中，参数 String 指明了窗口的标题。本题可以使用这个类的一些方法为这个实例设置相关的属性，例如，框架的大小、框架的背景色等。

上面创建的这个实例是不可见的，因此还需要调用 setVisible(true)方法来显示这个框架。

注意：在这个例子中，没有增加对这个框架进行控制的代码。为了能正常关闭并退出程序，这里使用了 System.exit(0)方法。当用户确认程序的执行结果是正确的之后，可以在 DOS 窗口中使用回车键退出程序。

程序代码实现如下：

```java
import java.awt.*;
import javax.swing.*;
import java.io.*;

public class MyFrame
{   public static void main(String args[])
    {   JFrame frame=new JFrame("My Frame");        //创建一个 JFrame 的实例
        frame.setSize(280, 300);                    //设置 JFrame 的大小
        frame.getContentPane().setBackground(Color.RED);
                                                    //设置 JFrame 的背景颜色
        frame.setVisible(true);                     //显示 JFrame

        BufferedReader intemp=new BufferedReader(
            new InputStreamReader(System.in));
        System.out.println("Press return key to exit.");
        try
        {   String s=intemp.readLine();             //等待用户的输入以关闭窗口
        } catch (IOException e )
        {   System.out.println("IOException");
        }
        System.exit(0);                             //强行退出并关闭
```

        }
    }

程序的执行结果如图 8-1 所示。

8.2 在习题 8.1 的 Frame 中增加一个背景为黄色的 Panel。

图 8-1　**MyFrame** 的执行结果

解：根据题意，要在前例生成的框架中，添加一个面板 JPanel。

Java 的图形用户界面由组件构成。显示在屏幕上的所有组件都必须包含在某个容器中，有些容器是可以嵌套的。习题 8.1 中创建的 JFrame 就是一种顶层容器。

而本题中要添加的面板（JPanel）是一种用途广泛的容器。与顶层容器不同的是，面板不能独立存在，必须被添加到其他容器内部。这就需要使用布局管理器来安排它们的位置。

本例中使用了一个空的布局管理器，面板位于框架的左上角。读者可以试着在框架中添加其他的布局管理器，并将面板置于特定的位置上。读者也可以试着使用框架类的一些方法来定制这个框架，比如用 setResizable(false) 方法让框架的大小不能改变等。

程序代码实现如下：

```java
import java.awt.*;
import javax.swing.*;
import java.io.*;

public class FrameWithPanelTest
{   public static void main(String args[])
    {   JFrame frame=new JFrame("My Frame");        //创建一个 JFrame 的实例
        frame.setSize(280, 300);                     //设置 JFrame 的大小
        frame.getContentPane().setBackground(Color.RED);
                                                      //设置 JFrame 的背景颜色
        frame.setVisible(true);                       //显示 JFrame

        JPanel contentPane=new JPanel();              //创建一个 JPanel 的实例
        contentPane.setSize(100, 100);                //设置 JPanel 的大小
        contentPane.setBackground(Color.yellow);      //设置 JPanel 的背景颜色
        frame.add(contentPane);                       //将 JPanel 添加到 JFrame 中

        BufferedReader intemp=new BufferedReader(
            new InputStreamReader(System.in));
        System.out.println("Press return key to exit.");
        try
        {   String s=intemp.readLine();              //等待用户的输入以关闭窗口
        } catch (IOException e )
        {   System.out.println("IOException");
        }
```

```
        System.exit(0);                              //强行退出并关闭
    }
}
```

程序的执行结果如图 8-2 所示。

8.3 在习题 8.2 的 Panel 中加入 3 个按钮,按钮上分别显示"打开""关闭""返回",并在一行内排开。

解:在前例生成的带面板的框架中,再添加 3 个按钮。因为要求按钮在一行内排开,因此必须使用布局管理器。此处使用 FlowLayout 布局管理器,3 个按钮并排置于面板中,面板居中置于框架内。

图 8-2 FrameWithPanelTest 的执行结果

程序代码实现如下:

```
import java.awt.*;
import javax.swing.*;
import java.io.*;

public class ThreeButtonsFrameTest
{   public static void main(String args[])
    {   JFrame frame=new JFrame("My Frame");        //创建一个 JFrame 的实例
        frame.setSize(400,500);                      //设置 JFrame 的大小
        frame.getContentPane().setBackground(Color.RED);
                                                     //设置 JFrame 的背景颜色
        //设置 JFrame 的布局管理器为 FlowLayout
        frame.setLayout(new FlowLayout(FlowLayout.CENTER, 50, 50));
        JPanel contentPane=new JPanel();             //创建一个 JPanel 的实例
        contentPane.setSize(100, 100);               //设置 JPanel 的大小
        //设置 JPanel 的布局管理器为 FlowLayout
        contentPane.setLayout(new FlowLayout());
        contentPane.setBackground(Color.yellow);     //设置 JPanel 的背景颜色

        JButton btn1, btn2, btn3;                    //定义 3 个按钮
        btn1=new JButton("打开");                    //定义按钮上的标识
        btn2=new JButton("关闭");
        btn3=new JButton("返回");
        contentPane.add(btn1);                       //将按钮添加到面板 JPanel 中
        contentPane.add(btn2);                       //使用 JPanel 的布局管理器
        contentPane.add(btn3);

        //将面板 JPanel 添加到 JFrame 中,使用 JFrame 的布局管理器
        frame.add(contentPane);
        frame.setVisible(true);                      //显示 JFrame

        BufferedReader intemp=new BufferedReader(
```

```
            new InputStreamReader(System.in));
    System.out.println("Press return key to exit.");
    try
    {   String s=intemp.readLine();              //等待用户的输入以关闭窗口
    } catch (IOException e )
    {   System.out.println("IOException");
    }
    System.exit(0);                              //强行退出并关闭
  }
}
```

程序的执行结果如图 8-3 所示。

8.4 Java 中提供了几种布局管理器？简述它们之间的区别。

图 8-3 ThreeButtonsFrameTest 的执行结果

解：用户程序界面的设计不仅要求其功能完备、设计合理，更要求界面风格赏心悦目，以人为本。为了帮助程序员设计良好的界面，减轻他们的工作强度，Java 提供了布局管理器，用来设计和控制各种组件在界面中的位置和相互关系。布局管理器接口 LayoutManager 定义在 java.awt 包中，在这个包中还提供了预定义的布局管理器类。

Java 平台提供了多种布局管理器，例如 java.awt.FlowLayout、java.awt.BorderLayout、java.awt.GridLayout、java.awt.GridBagLayout、java.awt.CardLayout、javax.swing.BoxLayout 和 javax.swing.SpringLayout 等。下面对其中几个较常用的布局管理器进行介绍。

(1) FlowLayout 型布局管理器

FlowLayout 型布局管理器是 JPanel 和 Applet 默认使用的布局管理器。它将组件自左至右逐个排列在容器中的一行上，一行放满后就另起一个新行。这种布局管理器非常简单，它并不强行设定组件的大小，而是允许组件拥有它们自己所希望的尺寸。不足之处是：当各组件大小不一，且组件数目比较多时，它的布局显得有点凌乱。

FlowLayout 有三种构造方法，分别是：

```
public FlowLayout()
public FlowLayout(int align)
public FlowLayout(int align, int hgap, int vgap)
```

在默认情况下，FlowLayout 将组件居中放置在容器的某一行上，当然也可以通过对齐方式选项 align，将组件的对齐方式设定为左对齐或者右对齐。FlowLayout 的构造方法中还有一对可选项 hgap 和 vgap，使用这对可选项可以设定组件之间的水平间距和垂直间距。

(2) BorderLayout 型布局管理器

BorderLayout 型布局管理器也是一种简单的布局管理器，它是 JDialog 类和 JFrame

类的默认布局管理器。它把要管理的容器划分成东(East)、南(South)、西(West)、北(North)、中(Center)5个区域。

BorderLayout 布局管理器有两种构造方法：
- BorderLayout()
- BorderLayout(int hgap, int vgap)

使用 BorderLayout 布局管理器，每个区域只能加入一个组件。如果试图向某个区域中加入多个组件，则必须使用容器的嵌套，保证最外层的组件是一个，否则区域中只有最后加入的一个组件是可见的。如果没有指定 hgap 和 vgap，则东、南、西、北、中各部分间的间距为 0，否则按指定的 hgap 和 vgap 留出空隙。如果在 add() 方法中没有指定将组件放到哪个区域，那么组件将被放置在 Center 区域。如果东、南、西、北、中 5 个区域中的某个区域没有使用，那么它的大小将变为零，由中心区域扩展并占据未使用的位置。

(3) GridLayout 型布局管理器

GridLayout 型布局管理器是使用较多的一种布局管理器，它将容器空间划分成若干行乘若干列的网格。组件按行依次放入这些划分出来的小格中，每个组件占据一格。一行放完后，再放到下一行中。每个网格中都必须放入组件，如果想要某个网格为空，可以使用空标签。

这种布局管理器划分的行数和列数完全由程序员自由控制，相应地，组件的定位也非常准确。但布局管理器中所有网格的大小一样，放入网格中的组件大小也相同，并且要求按固定的次序将组件放入网格中。

GridLayout 有三种构造方法：
- public GridLayout()
- public GridLayout(int rows, int cols)
- public GridLayout(int rows, int cols, int hgap, int vgap)

使用 GridLayout 的步骤为：

① 为容器设置布局管理器，并指定划分网格的行数 rows 和列数 cols：

```
setLayout(new GridLayout(rows, cols));
```

② 调用容器的 add() 方法加入相应的组件。

(4) CardLayout 型布局管理器

CardLayout 型布局管理器是一种卡片式的布局管理器，它将容器中的组件处理为一系列卡片。因为卡片的数量可以很多，所以它可以容纳多个组件。但每一时刻只显示出其中的一张，它占据了所有的容器空间。

使用 CardLayout 的步骤为：

① 为容器设置布局管理器：

```
setLayout(new CardLayout())
```

② 调用容器的 add() 方法加入相关的组件，并为组件指定各自的名字。

③ 调用布局管理器 CardLayout 的 show() 方法，显示所需的组件。

(5) GridBagLayout 型布局管理器

java.awt 还提供了 GridBagLayout 布局管理器,它是在 GridLayout 基础上发展而来的。GridLayout 中,网格和组件大小要求一致,组件的放入顺序相对固定,很不灵活。对于某些应用来说,这些要求过于苛刻。在 GridBagLayout 中,可以为每个组件指定其包含的网格数,允许保持组件原来的大小,并可以以任意次序将组件加入到容器中。组件既可以占多个网格,也可以只占一个网格的一部分,其组件的布局和加入顺序自由灵活。

(6) BoxLayout 型布局管理器

BoxLayout 布局管理器是定义在 javax.swing 包中的布局管理器,它将容器中的组件按水平方向排成一行或按垂直方向排成一列。当组件排成一行时,每个组件可以有不同的宽度;当组件排成一列时,每个组件可以有不同的高度。

BoxLayout 构造方法的格式为:

```
BoxLayout(Container target,int axis)
```

其中,Container 型参数 target 指明是为哪个容器设置此 BoxLayout 布局管理器的;int 型参数 axis 指明组件的排列方向,通常使用的是常量 BoxLayout.X_AXIS 或 BoxLayout.Y_AXIS,分别表示按水平方向排列和按垂直方向排列。

【拓展思考】

(1) 什么时候布局管理器要重新布置?

解:当布局管理器所管理的组件的显示受到影响时,例如改变容器的大小或在容器中添加了新的组件时,布局管理器都要重新布置组件的排列。

(2) FlowLayout 如何操作?

解:FlowLayout 在一行中放尽量多的组件。必要时可将组件放到多行中。

(3) 在 BoxLayout 中的 glue 组件有什么效果?

解:BoxLayout 中的 glue 组件管理布局管理器中的剩余空间。有空间时,它会扩展;但如果没有额外空间的话,它就不占地方。

8.5 BorderLayout 布局管理器是如何安排组件的?

解:BorderLayout 是顶层容器中内容窗格的默认布局管理器,它提供了一种较为复杂的组件布局管理方案,每个由 BorderLayout 管理的容器被划分成北(North)、南(South)、西(West)、东(East)、中(Center)5 个区域,分别代表容器的上、下、左、右和中部。这 5 部分分别用常量 BorderLayout.NORTH、BorderLayout.SOUTH、BorderLayout.WEST、BorderLayout.EAST 和 BorderLayout.CENTER 来表示。

在 BorderLayout 布局管理器的管理下,组件必须通过 add() 方法加入到容器中的指定区域中。如果在 add() 方法中没有指定将组件放到哪个区域,那么系统默认将它放置在 Center 区域,例如:

```
frame.getContentPane().add(button);
```

按钮将被放在框架的中部。

在容器的每个区域,只能加入一个组件。如果试图向某个区域中加入多个组件,那么只有最后一个组件是有效的,也就是只显示最后一个组件。当需要向一个区域中放置多个组件时,就需要利用容器的嵌套,将多个组件先放置在一个容器中,再把这个容器当作唯一的组件放到目标区域中。

对 East、South、West 和 North 这 4 个边界区域,如果其中的某个区域没有使用,那么它的大小将变为零,此时 Center 区域将会扩展并占据这个未用区域的位置。如果 4 个边界区域均没有使用,那么 Center 区域将会占据整个窗口。

8.6 如果想按上、中、下的位置安排 3 个按钮,可以使用哪些布局管理器? 编写程序实现之。

解:首先可以使用 BorderLayout 布局管理器,分别将 3 个按钮置于窗体的 North、Center 和 South 3 个区域,排成一列。

程序代码实现如下:

```
import java.awt.*;
import javax.swing.*;
import java.io.*;

public class ThreeButtonsJFrameTest
{   private JFrame frame;
    private JButton btn1,btn2,btn3;                    //定义 3 个按钮

    public static void main(String args[])
    {   ThreeButtonsJFrameTest that=new ThreeButtonsJFrameTest();
        that.go();
        BufferedReader intemp=new BufferedReader(
            new InputStreamReader(System.in));
        System.out.println( "Press return key to exit." );
        try
        {   String s=intemp.readLine();
        } catch (IOException e ){
            System.out.println( "IOException" );
        }
        System.exit(0);
    }

    void go()
    {   frame=new JFrame("Border Layout");
        btn1=new JButton("North");
        btn2=new JButton("Center");
        btn3=new JButton("South");

        frame.getContentPane().add(btn1, BorderLayout.NORTH);
                                                       //添加按钮 btn1 到北部
        frame.getContentPane().add(btn2, BorderLayout.CENTER);
                                                       //添加按钮 btn2 到中部
```

```
            frame.getContentPane().add(btn3, BorderLayout.SOUTH);
                                            //添加按钮 btn3 到南部

            //frame.setSize(350, 200);       //可以定制 Frame 的大小
            frame.pack();                    //也可以使用预定的尺寸
            frame.setVisible(true);
        }
    }
```

程序的执行结果如图 8-4 所示。

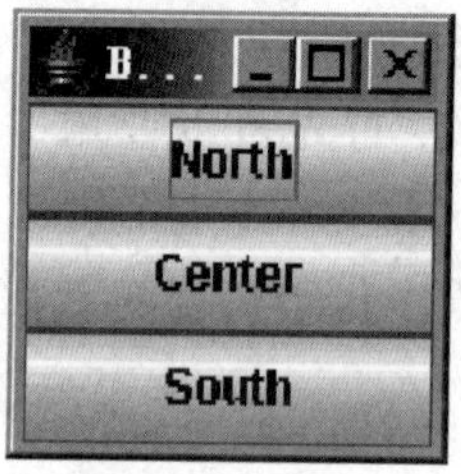

此外，还可以使用 GridLayout 来实现。将该布局管理器设置为三行一列，每行添加一个按钮。

程序代码实现如下：

图 8-4 ThreeButtonsJFrameTest 的执行结果

```
import java.awt.*;
import javax.swing.*;
import java.io.*;

public class ThreeButtonsJFrameTest2
{   private JFrame frame;
    private JButton btn1,btn2,btn3;             //定义 3 个按钮

    public static void main(String args[])
    {   ThreeButtonsJFrameTest2 that=new ThreeButtonsJFrameTest2();
        that.go();
        BufferedReader intemp=new BufferedReader(
            new InputStreamReader(System.in));
        System.out.println( "Press return key to exit." );
        try
        {   String s=intemp.readLine();
        } catch (IOException e)
        {   System.out.println( "IOException" );
        }
        System.exit(0);
    }

    void go()
    {   frame=new JFrame("Grid Layout");

        btn1=new JButton("Button1");
        btn2=new JButton("Button2");
        btn3=new JButton("Button3");

        //设置 GridLayout 布局管理器为三行一列
```

```
        frame.getContentPane().setLayout(new GridLayout(3,1));
        frame.getContentPane().add(btn1);            //添加按钮 btn1
        frame.getContentPane().add(btn2);            //添加按钮 btn2
        frame.getContentPane().add(btn3);            //添加按钮 btn3

        frame.pack();
        frame.setVisible(true);
    }
}
```

程序的执行结果如图 8-5 所示。

8.7 Frame 和 Panel 默认的布局管理器分别是什么类型？

图 8-5 ThreeButtonsJFrameTest2 的执行结果

解：Frame 类默认的布局管理器是 BorderLayout，Panel 类默认的布局管理器是 FlowLayout。这两类布局管理器的介绍请参见主教材及本章的习题 8.4。

8.8 将习题 8.2 中 Panel 的布局管理器设置为 GridLayout，然后向其中加入 3 个按钮。

解：GridLayout 有三种构造方法，可以自行设定网格的个数及网格之间的间隔。要添加 3 个按钮，可以设定布局管理器中的行数为 1，列数为 3，网格之间无间隔，相应的设置如下：

```
contentPane.setLayout(new GridLayout(1,3));
```

程序代码实现如下：

```
import java.awt.*;
import javax.swing.*;
import java.io.*;

public class PanelWithLayoutTest
{   public static void main(String args[])
    {   JFrame frame=new JFrame("My Frame");              //创建一个 JFrame 的实例
        frame.setSize(280,300);                           //设置 JFrame 的大小
        frame.getContentPane().setBackground(Color.RED);
                                                          //设置 JFrame 的背景颜色
        //设置布局管理器
        frame.setLayout(new FlowLayout(FlowLayout.CENTER, 50, 100));

        JPanel contentPane=new JPanel();                  //创建一个 JPanel 的实例
        contentPane.setSize(100, 100);                    //设置 JPanel 的大小
        contentPane.setLayout(new GridLayout(1,3));//设置面板的布局管理器
        contentPane.setBackground(Color.yellow);          //设置 Jpanel 的背景颜色
```

```
JButton btn1, btn2, btn3;              //定义 3 个按钮
btn1=new JButton("打开");               //定义按钮标识
btn2=new JButton("关闭");
btn3=new JButton("返回");
contentPane.add(btn1);                  //将按钮添加到面板中
contentPane.add(btn2);
contentPane.add(btn3);

frame.add(contentPane);                 //将面板 contentPane 添加到 JFrame 中
frame.setVisible(true);                 //显示 JFrame

BufferedReader intemp=new BufferedReader(
    new InputStreamReader(System.in));
System.out.println("Press return key to exit.");
try
{   String s=intemp.readLine();
} catch (IOException e )
{   System.out.println("IOException");
}
System.exit(0);
    }
}
```

程序的执行结果如图 8-6 所示。

如果想让 3 个按钮之间留些空隙,则可以设定网格之间的间隔,例如 20 个像素。相应的设定语句为:

```
contentPane.setLayout(new GridLayout(1, 3, 20, 20));
```

程序的执行结果如图 8-7 所示。

图 8-6　GridLayout 示例之一　　　　　　图 8-7　GridLayout 示例之二

此外,还可以多设网格的列数,并将按钮分别放入不相邻的列中,在空白的网格中加入一个空标签。例如,设定 5 列网格,并将 3 个按钮分别放入第 1、3、5 列中,在第 2、4 列放入空标签。

程序修改为:

```
import java.awt.*;
import javax.swing.*;
import java.io.*;
```

```java
public class PanelWithLayoutTest
{   public static void main(String args[])
    {   JFrame frame=new JFrame("My Frame");        //创建一个 JFrame 的实例
        frame.setSize(280,300);                     //设置 JFrame 的大小
        frame.getContentPane().setBackground(Color.RED);
                                                    //设置 JFrame 的背景颜色
        //设置布局管理器
        frame.setLayout(new FlowLayout(FlowLayout.CENTER, 50, 100));

        JPanel contentPane=new JPanel();            //创建一个 JPanel 的实例
        contentPane.setSize(100, 100);
        //contentPane.setLayout(new GridLayout(1,3));
        contentPane.setLayout(new GridLayout(1, 5));   //设置面板的布局管理器
        contentPane.setBackground(Color.yellow);

        JButton btn1,btn2,btn3;                     //定义 3 个按钮
        btn1=new JButton("打开");
        btn2=new JButton("关闭");
        btn3=new JButton("返回");
        contentPane.add(btn1);                      //将按钮添加到面板 contentPane 中
        contentPane.add(new JLabel());              //将空标签添加到面板中,占位
        contentPane.add(btn2);
        contentPane.add(new JLabel());
        contentPane.add(btn3);

        frame.add(contentPane);                     //将面板 contentPane 添加到 JFrame 中
        frame.setVisible(true);                     //显示 JFrame

        BufferedReader intemp=new BufferedReader(
            new InputStreamReader(System.in));
        System.out.println("Press return key to exit.");
        try
        {   String s=intemp.readLine();
        } catch (IOException e )
        {   System.out.println("IOException");
        }
        System.exit(0);
    }
}
```

修改后程序的执行结果如图 8-8 所示。

8.9 将习题 8.2 中 Panel 的布局管理器设置为 BorderLayout,然后分别向其中的每个区域加入

图 8-8　GridLayout 示例之三

一个按钮。

解：参看习题 8.5 的解答。BorderLayout 是顶层容器中内容窗格的默认布局管理器，它管理的每个容器被划分成北（North）、南（South）、西（West）、东（East）、中（Center）5 个区域，分别代表容器的上、下、左、右和中部。这 5 部分分别用常量 BorderLayout. NORTH、BorderLayout. SOUTH、BorderLayout. WEST、BorderLayout. EAST 和 BorderLayout. CENTER 表示。在容器的每个区域，可以加入一个组件。

根据题意，定义一个面板 JPanel，并使用 BorderLayout 布局。为 JPanel 的 5 个区域分别定义各自的按钮，在每个区域中添加一个按钮。

程序代码实现如下：

```java
import java.awt.*;
import javax.swing.*;
import java.io.*;

public class PanelWithLayoutTest2
{   public static void main(String args[])
    {   JFrame frame=new JFrame("My Frame");           //创建一个 JFrame 的实例
        frame.setSize(280, 300);                        //设置 JFrame 的大小
        frame.getContentPane().setBackground(Color.RED);
                                                        //设置 JFrame 的背景颜色
        //设置 JFrame 的布局管理器
        frame.setLayout(new FlowLayout(FlowLayout.CENTER, 50, 100));
        JPanel contentPane=new JPanel();               //创建一个 JPanel 的实例
        contentPane.setSize(100, 100);                  //设置 JPanel 的大小
        contentPane.setLayout(new BorderLayout());      //设置 JPanel 的布局管理器
        contentPane.setBackground(Color.yellow);        //设置 JPanel 的背景颜色

        JButton btn1, btn2, btn3, btn4, btn5;           //定义 5 个按钮
        btn1=new JButton("SOUTH");                      //按钮的标签
        btn2=new JButton("NORTH");
        btn3=new JButton("EAST");
        btn4=new JButton("WEST");
        btn5=new JButton("CENTER");

        contentPane.add(btn1,BorderLayout.SOUTH);       //将按钮添加到面板中
        contentPane.add(btn2,BorderLayout.NORTH);
        contentPane.add(btn3,BorderLayout.EAST);
        contentPane.add(btn4,BorderLayout.WEST);
        contentPane.add(btn5,BorderLayout.CENTER);

        frame.add(contentPane);                         //将面板 contentPane 添加到 JFrame 中
        frame.setVisible(true);                         //显示 JFrame
```

```
        BufferedReader intemp=new BufferedReader(
            new InputStreamReader(System.in));
        System.out.println("Press return key to exit.");
        try
        {   String s=intemp.readLine();
        } catch (IOException e )
        {   System.out.println("IOException");
        }
        System.exit(0);
    }
}
```

程序的执行结果如图 8-9 所示。

8.10 什么是事件？事件是怎样产生的？

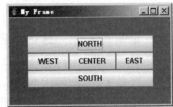

图 8-9　BorderLayout 示例

解：图形界面是一种非常灵活、直观的人机接口，通过移动鼠标和使用键盘，就可以对界面进行操作及控制，实现人机交互。在 Java 程序运行过程中，每当使用者通过用户界面进行某个操作时，便会引发一个相应的事件。为了能够接收用户的命令，系统预定义了一些事件，用来识别用户的鼠标及键盘操作，并按照用户的程序代码做出响应。实际上事件是一个描述用户所执行操作的数据对象。

Java 中，除了键盘和鼠标操作外，系统状态的改变、标准图形界面元素等都可能引发事件。图形用户界面中可能产生事件的每个元素被称为事件源，例如界面组件、窗口和菜单等，不同事件源上发生的事件的种类不同。事件的来源是用户的操作，每当用户在组件上进行某种操作，例如用鼠标单击按钮、键盘输入某个符号时，系统的事件处理部分便会生成一个事件对象。事件对象描述的是用户所执行的操作，用户的操作不同，事件对象的内容也会不同。

【拓展思考】

(1) 什么是鼠标事件？

解：用户以各种不同方式操纵鼠标时生成的事件就是鼠标事件。具体到某一种情形，只会关注不同类的鼠标事件，包括移动鼠标、按下鼠标按钮、鼠标进入特定组件所在区域及拖拽鼠标等。

(2) 什么是键盘事件？

解：按下键盘键时生成的事件就是键盘事件，它能让监听器程序立即响应用户的输入。代表事件的对象中保存的编码可以指明按下的是哪个键。

(3) 什么是事件适配器类？

解：事件适配器类是实现了监听器接口的一个类，它的所有方法都只有空定义。继承相应的适配器类，并只需实现其中感兴趣的方法，也可以创建监听器类。

8.11 在 API 文档中查找 Event 类，解释其中 target、when、id、x、y 和 arg 分别表示什么内容？

解：Event 是 java.awt 中定义的一个类，target、when、id、x、y 和 arg 都是其中的主要

成员变量,其中:
- target 是 Object 类型,它表示此事件涉及的组件,即用户操作的对象。
- when 是 long 类型,它表示产生此事件的时间。
- id 是 int 类型,它表示事件的类型。
- x 和 y 都是 int 类型,分别表示产生事件的横、纵坐标位置。
- 最后一个 arg 是 Object 类型,它是与具体事件有关的参数。

参数 id 表示的事件类型包括:
- 窗口事件,例如移动窗口、关闭窗口等。
- 鼠标事件,例如移动鼠标,按下或松开鼠标键等。
- 焦点事件,例如获得焦点或失去焦点。
- 键盘事件,例如按下或松开鼠标键。
- ACTION 事件,例如按动图形用户界面中的按钮。
- 列表框事件,例如在列表框中进行选择。
- 滚动条事件,例如按动滚动条中的向上或向下按钮。

8.12 设计鼠标控制程序。程序运行时,如果在窗口中移动鼠标,则窗口底部将显示出鼠标的当前位置。如果移动鼠标的同时还按住 Ctrl 或 Shift 键,则窗口底部还会显示出 C 或 S。如果按下键盘上的键,程序窗口的底部将显示出字母"D";如果松开键盘上的键,程序窗口的底部将显示出字母"U"。

解:由题可知,该程序需要同时实现 MouseMotionListener、KeyListener 两个接口,并对窗体注册相应的监听程序。对于鼠标和键盘事件的响应,可以由 JLabel 显示内容来查看。

程序代码实现如下:

```
import java.awt.*;
import java.awt.event.*;
import javax.swing.*;

public class MouseControl implements MouseMotionListener, KeyListener {
    //MouseControl 类同时实现 MouseMotionListener、KeyListener 两个接口
    private JFrame frame;
    private JLabel tf;
    String ch="";                                       //放置待显示的字符

    public static void main(String args[])
    {   MouseControl two=new MouseControl();
        two.go();
    }

    public void go()
    {   frame=new JFrame("Mouse Control ");
        Container contentPane=frame.getContentPane();
```

```java
        contentPane.add(new JLabel ("get mouse and keyboard event"),
    BorderLayout.NORTH);
    tf=new JLabel();

        contentPane.add(tf,BorderLayout.SOUTH);

        //注册监听程序
        frame.addMouseMotionListener(this);
        frame.addKeyListener(this);

        frame.setSize(300,300);
        frame.setVisible(true);
    }

    //实现 MouseMotionListener 接口中的方法
    public void mouseDragged (MouseEvent e)
    {   String s="Mouse dragging: X="+e.getX()+"Y="+e.getY()+" key: "+ch;
        tf.setText(s);
    }

    public void mouseMoved (MouseEvent e)
    {   String s="Mouse dragging:   X="+e.getX()+"Y="+e.getY()+" key: "+ch;
        tf.setText(s);
    }

    //实现 KeyListener 接口中的方法
    public void keyTyped(KeyEvent evt)
    {   int charcode=evt.getKeyCode();
        ch="D";
        if (charcode==KeyEvent.VK_SHIFT) ch="S";
        if (charcode==KeyEvent.VK_CONTROL) ch="C";
    }
    public void keyPressed(KeyEvent evt)
    {   int charcode=evt.getKeyCode();
        ch="D";                                             //其他键显示"D"
        if (charcode==KeyEvent.VK_SHIFT) ch="S";     //Shift 显示"S"
        if (charcode==KeyEvent.VK_CONTROL) ch="C"; //Ctrl 显示"C"
    }

    public void keyReleased(KeyEvent evt)           //松开时显示"U"
    {   ch="U";
    }
}
```

程序的执行结果如图 8-10 所示。

8.13 委托事件处理模型是怎样对事件进行处理的？事件监听程序的作用是什么？

图 8-10 MouseControl 类的执行结果

解：委托事件处理模型是在改进层次事件处理模型之后提出的一种新的事件处理方式。在委托事件处理模型中，用户操作引发的事件对象仍然传递给相应组件。为了接收事件对象并进行事件处理，组件必须注册一个事件处理程序，也就是事件的监听程序（Listener）。事件的监听程序可以定义在组件所在的类中，也可以定义在其他的类里；而对事件的处理，则由组件委托给事件监听程序所在的类来完成。

事件源是一种对象，它可以注册监听程序对象，事件源向已经注册的所有监听程序发送事件对象，并由监听程序来处理所发生的事件。

在委托事件处理模型下，事件对象只被传递给已经注册的监听程序。监听程序对象根据事件对象内封装的信息，决定如何响应这个事件。根据用户操作的不同，事件也被分成不同类型，例如窗口事件、键盘事件、鼠标事件等。每种事件都有一个对应的监听程序类，这个监听程序类应该实现相应的监听程序接口（Listener Interface），并且定义了事件对象的接收和处理方法。

8.14 java.awt.event 中定义了哪些事件类？各类对应的接口是什么？各接口中都声明了哪些方法？

解：java.awt.event 中定义了如下的事件类：

ActionEvent	AdjustmentEvent	AWTEventListenerProxy
ComponentAdapter	ComponentEvent	ContainerAdapter
ContainerEvent	FocusAdapter	FocusEvent
HierarchyBoundsAdapter	HierarchyEvent	InputEvent
InputMethodEvent	InvocationEvent	ItemEvent
KeyAdapter	KeyEvent	MouseAdapter
MouseEvent	MouseMotionAdapter	MouseWheelEvent
PaintEvent	TextEvent	WindowAdapter
WindowEvent		

事件类中定义的接口及接口中声明的相应方法见表 8-1 所示。

表 8-1 事件类中定义的接口及方法

事　件　类	接　　口	方　　法
ActionEvent	ActionListener	actionPerformed(ActionEvent e)
AdjustmentEvent	AdjustmentListener	adjustmentValueChanged(AdjustmentEvent e)
ComponentEvent	ComponentListener	componentHidden(ComponentEvent e) componentMoved(ComponentEvent e) componentResized(ComponentEvent e) componentShown(ComponentEvent e)

续表

事 件 类	接 口	方 法
ContainerEvent	ContainerListener	componentAdded(ContainerEvent e) componentRemoved(ContainerEvent e)
FocusEvent	FocusListener	focusGained(FocusEvent e) focusLost(FocusEvent e)
HierarchyEvent	HierarchyBoundsListener	ancestorMoved(HierarchyEvent e) ancestorResized(HierarchyEvent e)
	HierarchyListener	hierarchyChanged(HierarchyEvent e)
InputMethodEvent	InputMethodListener	caretPositionChanged(InputMethodEvent event) inputMethodTextChanged(InputMethodEvent event)
ItemEvent	ItemListener	itemStateChanged(ItemEvent e)
KeyEvent	KeyListener	keyPressed(KeyEvent e) keyReleased(KeyEvent e) keyTyped(KeyEvent e)
MouseEvent	MouseListener	mouseClicked(MouseEvent e) mouseEntered(MouseEvent e) mouseExited(MouseEvent e) mousePressed(MouseEvent e) mouseReleased(MouseEvent e)
	MouseMotionListener	mouseDragged(MouseEvent e) mouseMoved(MouseEvent e)
MouseWheelEvent	MouseWheelListener	mouseWheelMoved(MouseWheelEvent e)
TextEvent	TextListener	textValueChanged(TextEvent e)
WindowEvent	WindowFocusListener	windowGainedFocus(WindowEvent e) windowLostFocus(WindowEvent e)
	WindowListener	windowActivated(WindowEvent e) windowClosed(WindowEvent e) windowClosing(WindowEvent e) windowDeactivated(WindowEvent e) windowDeiconified(WindowEvent e) windowIconified(WindowEvent e) windowOpened(WindowEvent e)
	WindowStateListener	windowStateChanged(WindowEvent e)

第 9 章 Swing 组件

9.1 编写一个程序,使之具有如主教材中图 9-26 所示的界面,并实现简单的控制:按 Clear 按钮时清空两个文本框的内容,按 Copy 按钮时将 Source 文本框的内容复制到 Target 文本框,按 Close 按钮则结束程序的运行。

解:图 9-1 所示的窗口中定义了两个标签(分别是 Source 和 Target)、两个文本组件和 3 个按钮(分别是 Clear、Copy 和 Close)。窗口是框架 Frame 类的。为使各组件的布局有序,此处使用布局管理器

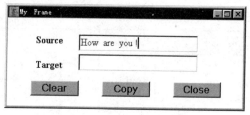

图 9-1 习题 9.1 的目标窗口

GridLayout,将标签 Source 及对应的一个文本组件先组织到一个面板中,再放入 GridLayout 的一个网格中,同样处理标签 Target 及与它同一行的文本组件。将 3 个按钮放到第 3 个面板中,并放入新的网格中。在标签 Target 所在的面板和 3 个按钮所在的面板中间插入一个空标签,使得它们之间留有适当的空间。3 个按钮所在的面板中,设定横向间距为 40,保证 3 个按钮等间距分布。

对 3 个按钮 Clear、Copy 和 Close 分别添加监听程序,并实现 ActionListener 接口中的 actionPerformed() 方法。当单击按钮时,监听程序将 ActionEvent 类型的一个实例 e 传入 actionPerformed() 方法,方法中判断按钮是哪一个,并针对按钮做出响应。

使用 String buttonName = e.getActionCommand() 就可以知道按下的是哪个按钮。通过 Button.getLabel() 方法可得到各个按钮的名称,再与 buttonName 比较,进而对某个按钮进行相应的控制。

也可以使用 e.getSource() 得到事件源,判断事件源的按钮名,进而进行相应的处理。例如当事件源是按钮 bClear 时,使用 tfs.setText(sBlank),清空相应的文本框内的内容。当事件源是 bCopy 时,先用 tfs.getText() 获得第一个文本框内的内容,再写入第二个文本框中。

程序代码实现如下:

```
import java.awt.*;
import java.awt.event.*;
import javax.swing.*;

public class CopyText extends WindowAdapter implements ActionListener
{   JFrame f;
    JButton bClear, bCopy, bClose;              //定义 3 个按钮
```

```java
    JLabel lSource, lTarget, lSpace1, lSpace2;   //定义两个标签及两个占位用空标签
    JTextField tfs, tft;                          //定义两个文本组件
    String ad="How are you!";

    public static void main(String args [])
    {   CopyText be=new CopyText();
        be.go();
    }

    public void go()
    {   f=new JFrame("My JFrame");                            //定义框架
        f.setLayout(new GridLayout(6, 1, 10, 10));            //设置布局管理器
        f.setSize(350, 250);                                  //设置框架的尺寸

        lSource=new JLabel("Source");                         //创建两个标签
        lTarget=new JLabel("Target");
        lSpace1=new JLabel();                                 //创建两个占位空标签
        lSpace2=new JLabel();
        tfs=new JTextField(ad, 15);                           //创建两个文本组件
        tft=new JTextField(15);
        JPanel pan1=new JPanel();                             //创建 3 个面板
        JPanel pan2=new JPanel();
        JPanel pan3=new JPanel();
        //面板 pan3 的布局管理器
        pan3.setLayout(new FlowLayout(FlowLayout.CENTER, 40, 0));
        pan1.setSize(100, 50);                                //面板 pan1 的尺寸
        pan2.setSize(100, 50);                                //面板 pan2 的尺寸
        pan3.setSize(100, 50);                                //面板 pan3 的尺寸
        pan1.add(lSource);                                    //将第一个标签加入面板 pan1 中
        pan1.add(tfs);                                        //将第一个文本组件加入面板 pan1 中
        pan2.add(lTarget);                                    //将第二个标签加入面板 pan2 中
        pan2.add(tft);                                        //将第二个文本组件加入面板 pan2 中

        bClear=new JButton("Clear");                          //创建 3 个按钮
        bCopy=new JButton("Copy");
        bClose=new JButton("Close");
        bClear.addActionListener(this);                       //建立按钮的监听程序
        bCopy.addActionListener(this);
        bClose.addActionListener(this);
        pan3.add(bClear);                                     //将 3 个按钮加入面板 pan3 中
        pan3.add(bCopy);
        pan3.add(bClose);

        f.add(lSpace1);                                       //将一个占位空标签加入框架内最上面的空白中
```

```
        f.add(pan1);                          //加入面板 pan1
        f.add(pan2);                          //加入面板 pan2
        //将一个占位空标签加入框架内三个按钮与文本组件间的空白中
        f.add(lSpace2);
        f.add(pan3);                          //将含 3 个按钮的面板加入框架中

        f.addWindowListener(this);            //对窗口的监听
        f.setVisible(true);                   //框架可见
    }

    //实现 ActionListener 接口中的 actionPerformed()方法
    public void actionPerformed(ActionEvent e)
    {    String sBlank="";

        //bClear 按钮的处理
        if (e.getSource()==bClear)
        {   tfs.setText(sBlank);              //清空文本组件内的内容(赋一个空串)
            tft.setText(sBlank);
        }
        //bCopy 按钮的处理
        if (e.getSource()==bCopy)
        {   //将第一个文本组件内的内容赋到第二个文本组件中
            tft.setText(tfs.getText());
        }
        //bClose 按钮的处理
        if (e.getSource()==bClose)
        {   System.exit(0);                   //关闭并退出
        }
    }
    public void windowClosing(WindowEvent e)
    {   System.exit(0);                       //退出
    }
}
```

9.2 编写一个程序,使之具有如主教材图 9-27 所示的界面。每当在右侧的选择框中选中一个人的名字时,便在左侧的文本区中显示出此人的情况介绍;按 Close 按钮时,结束程序的运行。

解:图 9-2 所示的界面分左右两部分,右侧列有若干人名,左侧为对应的信息框。根据题意可以分别使用两个面板 JPanel 来显示。左侧的面板 pan1 使用 FlowLoyout 布局管理器。因为右侧除有人名框外,还有

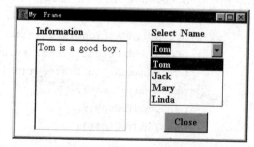

图 9-2 习题 9.2 的目标窗口

按钮，所以右侧的面板 pan2 使用 GridBagLayout 布局管理器。为了方便设计，程序中使用数组 classNameInfor 保存人名及其描述信息。人名列表的显示使用选择框 Choice，单击列表中的人名后，将对应的信息显示在左侧的文本区中。

与前一例相同，对按钮 Close 添加监听程序，并实现 ActionListener 接口中的 actionPerformed() 方法。

与按钮的监听程序不同，在选择框 Choice 上添加监听程序，则要实现 ItemListener 接口中的 itemStateChanged() 方法，而不是 ActionListener 接口中的 actionPerformed() 方法。监听程序中得到的是 ItemEvent 类型的实例 e，使用方法 e.getSource() 可以知道实例源是什么。由于我们知道这个实例源是 Choice 类型的，所以可以使用 Choice 中的方法得到所选名字的索引号，对应的方法是 Choice.getSelectedIndex()。按照索引可直接找到相应的信息，并将它显示在左侧的文本区中。

程序代码实现如下：

```java
import java.awt.*;
import java.awt.event.*;
import java.util.*;
import javax.swing.*;

class NameInformation
{   private String name, infor;
    public NameInformation(String n, String i)
    {   name=n;
        infor=i;
    }
    public String getName()
    {   return name;
    }
    public String getInfor()
    {   return infor;
    }
}

public class GetPeopleInfo extends WindowAdapter implements ActionListener, ItemListener
{   JFrame f;                                        //定义框架
    JButton bClose;                                  //定义按钮
    JLabel lInformation, lSelectName;                //定义两个标签
    JLabel lSpace1=new JLabel("");                   //定义 4 个占位空标签
    JLabel lSpace2=new JLabel("");
    JLabel lSpace3=new JLabel("");
    JLabel lSpace4=new JLabel("");
    JTextArea tTarget;                               //定义文本组件
    //人名及对应的信息
```

```java
NameInformation [] classNameInfor=new NameInformation[4];
JComboBox cName;                              //定义组合框(下拉菜单)

public static void main(String args [])
{   GetPeopleInfo be=new GetPeopleInfo();
    //初始化人名及对应的信息
    be.classNameInfor[0]=new NameInformation("Tom",
                        "Tom is a good boy.");
    be.classNameInfor[1]=new NameInformation("Jack",
                        "Jack is the best student of this class.");
    be.classNameInfor[2]=new NameInformation("Mary",
                        "Mary is a nice girl.");
    be.classNameInfor[3]=new NameInformation("Linda",
                        "Linda is the teacher of this class.");
    be.go();
}

public void go()
{   f=new JFrame( "My JFrame" );                  //创建框架
    f.setLayout(new GridLayout(1, 2));            //设置框架的布局管理器
    f.setSize(300, 250);                          //设置框架的尺寸

    lInformation=new JLabel( "Information" );     //创建两个标签
    lSelectName=new JLabel( "Select Name" );
    JPanel pan1=new JPanel();                     //创建两个面板
    JPanel pan2=new JPanel();

    pan1.setLayout(new FlowLayout(FlowLayout.LEFT)); //面板 pan1 的布局管理器
    GridBagLayout pan2Layout=new GridBagLayout();    //面板 pan2 的布局管理器
    pan2.setLayout(pan2Layout);
    GridBagConstraints c=new GridBagConstraints();   //对应于 GridBagLayout
    c.anchor=GridBagConstraints.NORTH;               //在网格中如何显示组件
    c.gridx=0;
    c.gridy=0;

    tTarget=new JTextArea("", 8, 10);                //创建文本组件
    //创建带滚卷条的显示区
    JScrollPane jsp1=new JScrollPane(tTarget,
        JScrollPane.VERTICAL_SCROLLBAR_ALWAYS,
        JScrollPane.HORIZONTAL_SCROLLBAR_ALWAYS);

    pan1.add(lInformation);                          //左侧面板中组件的加入
    pan1.add(jsp1);
```

```
        pan2Layout.setConstraints(lSelectName, c);      //右侧面板的组件放置方式
        pan2.add(lSelectName);

        cName=new JComboBox();                          //创建组合框
        for (int i=0; i<4; i++)
        {   cName.addItem(classNameInfor[i].getName()); //加入名字
        }
        cName.addItemListener(this);                    //添加监听器

        c.gridx=0;
        c.gridy=10;
        pan2Layout.setConstraints(cName, c);            //名字的显示方式
        pan2.add(cName, c);

        c.gridx=0;
        c.gridy=30;
        pan2Layout.setConstraints(lSpace1, c);
        pan2.add(lSpace1, c);

        c.gridx=0;
        c.gridy=50;
        pan2Layout.setConstraints(lSpace2, c);
        pan2.add(lSpace2, c);

        c.gridx=0;
        c.gridy=80;
        pan2Layout.setConstraints(lSpace3, c);
        pan2.add(lSpace3,c);

        c.gridx=0;
        c.gridy=100;
        pan2Layout.setConstraints(lSpace4, c);
        pan2.add(lSpace4, c);

        bClose=new JButton( "Close" );
        bClose.addActionListener(this);                 //监听

        c.gridx=0;
        c.gridy=170;
        c.anchor=GridBagConstraints.CENTER;
        pan2Layout.setConstraints(bClose, c);
        pan2.add(bClose, c);

        f.add(pan1);
```

```
            f.add(pan2);

            f.addWindowListener(this);                    //对窗口的监听
            f.setVisible(true);
        }

        //实现ActionListener接口中的actionPerformed()方法
        public void actionPerformed(ActionEvent e)
        {   //bClose JButton
            if (e.getSource()==bClose)
            {    System.exit(0);
            }
        }

        //实现ItemListener接口中的itemStateChanged()方法
        public void itemStateChanged(ItemEvent e)
        {   tTarget.replaceRange(classNameInfor
                [((JComboBox)(e.getSource())).getSelectedIndex()].getInfor(),
                0,tTarget.getText().length());
        }

        public void windowClosing(WindowEvent e)
        {   System.exit(0);
        }
    }
```

为简单起见,程序中直接写入了人名及对应的信息,读者可以修改这部分代码,从键盘读入人员数及相应的名和信息,并将之加入到人员列表中,扩充程序的功能,使之更加方便。

9.3 编写一个程序,使之具有如主教材图 9-28 所示的界面,并且实现计算器的基本功能。目标窗口如图 9-3 所示。

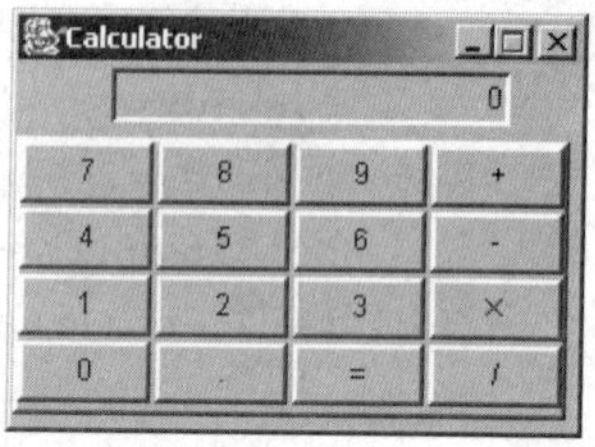

解:定义一个框架,在该框架中,定义两个面板 JPanel。其中,Panel1 存放运算结果,Panel2 存放 16 个 JButton 按钮。在 Panel1 中使用 JTextField 型的 tfResult 来显示结果,并设置为不可编辑。对于 Panel2 采用 4 行 4 列的

图 9-3 习题 9.3 的目标窗口

GridLayout 布局,然后使用 addButton 方法将数字按钮和运算符按钮添加到面板中,同时注册按钮事件监听器。事件处理方法 actionPerformed(ActionEvent e)主要完成对按钮事件的处理。事件的处理分数字、小数点、等号和运算符 4 种情况来处理。对于运算符,要调用 equalaction(ActionEvent e)方法进行计算,同时将结果显示在单行文本域中。

程序代码实现如下:
```
import java.awt.*;
```

```java
import javax.swing.*;
import java.awt.event.*;
import javax.swing.border.*;

public class Calculator1 extends JFrame implements ActionListener
{   private JPanel Panel1=new JPanel();
    private JPanel Panel2=new JPanel();
    private JTextField tfResult=new JTextField();

    private GridLayout gridLayout1=new GridLayout();
    private GridBagLayout gridBagLayout1=new GridBagLayout();
    private GridBagLayout gridBagLayout2=new GridBagLayout();

    //保存最近一次运算符(null:无,+:加,-:减,*:乘,/:除)
    private String recentOperation=null;
    //保存最近一次运算数据
    private String recentNum=null;
    //描述当前输入状态,即:是重新输入,还是接在后面
    private boolean isNew=true;

    public void addButton(Container c, String s)
    {   JButton b=new JButton(s);
        b.setFont(new java.awt.Font("SansSerif", 0, 12));
        b.setForeground(Color.black);
        b.setBorder(BorderFactory.createRaisedBevelBorder());
        c.add(b);
        b.addActionListener(this);                        //监听
    }

    public void actionPerformed(ActionEvent e)            //实现按钮动作的事件处理方法
    {   String s=e.getActionCommand();
        if(s.charAt(0)>='0' && s.charAt(0)<='9')          //数值
        {   if (!isNew)                                   //接续输入
                tfResult.setText(tfResult.getText()+s);   //拼接
            else                                          //重新输入
                tfResult.setText(s);                      //刷新
            isNew=false;                                  //改变标志
        }
        else if (s.equals("."))                           //输入的是小数点
        {       if (tfResult.getText().indexOf(".") !=-1) return;  //已有小数点
                if (!isNew && tfResult.getText() !="")    //拼接小数点
                    tfResult.setText(tfResult.getText()+".");      //有前导数字
                else
                    tfResult.setText("0.");                        //无前导数字,加0
```

```java
            isNew=false;
        }
        else if (s.equals("="))                              //输入等号=
        {   equalaction(e);                                  //进行计算
        }
        else                                                 //输入运算符
        {   if ((tfResult.getText()).equals("")) return;
            if (recentOperation!=null) equalaction(e);
            recentOperation=s;
            recentNum=tfResult.getText();
            isNew=true;
        }
    }

    void equalaction(ActionEvent e)
    {   if (recentOperation==null||recentNum==null||
            tfResult.getText().equals("")) return;           //没有输入
        double last=0 , now=0;

        try
        {   last=Double.parseDouble(recentNum);
            now=Double.parseDouble(tfResult.getText());
        } catch (NumberFormatException ne)
        {   recentOperation=null;
            recentNum=null;
            tfResult.setText("数据输入不合法");
            System.out.println("数据输入不合法。");
            isNew=true;
            return;
        }

        if (recentOperation.equals("+"))
        {   last +=now;
        }
        if (recentOperation.equals("-"))
        {   last -=now;
        }
        if (recentOperation.equals("*"))
        {   last *=now;
        }
        if (recentOperation.equals("/"))
        {   last /=now;
        }
        tfResult.setText(""+last);
```

```
        //tfResult.setText(tfResult.getText()+last);

        recentNum=tfResult.getText();
        recentOperation=null;
        isNew=true;
    }


    public Calculator1()
    {   //结果显示框的初始设置
        tfResult.setBorder(BorderFactory.createLoweredBevelBorder());
        tfResult.setDisabledTextColor(Color.white);
        tfResult.setEditable(false);
        tfResult.setText("0");
        tfResult.setHorizontalAlignment(SwingConstants.RIGHT);

        //++++++++++++++++++++++++++++++++++++++
        Panel1.setLayout(gridBagLayout1);
        Panel1.setPreferredSize(new Dimension(333, 30));
        Panel1.add(tfResult,new GridBagConstraints(1, 0, 1, 1, 0.0, 0.0,
            GridBagConstraints.CENTER, GridBagConstraints.NONE,
            new Insets(0, 15, 0, 0), 155, 0));

        //++++++++++++++++++++++++++++++++++++++
        Panel2.setBorder(BorderFactory.createRaisedBevelBorder());
        gridLayout1.setColumns(6);
        gridLayout1.setHgap(2);
        gridLayout1.setRows(4);
        gridLayout1.setVgap(2);
        Panel2.setLayout(gridLayout1);

        String buttons="789/456*123-0.=+";
        for (int i=0;i<buttons.length() ;i++)            //向面板中添加按钮
        {   addButton(Panel2,buttons.substring(i,i+1));
        }

        //+++++++++++++++++++++++++++++++++++++++
        this.getContentPane().setLayout(gridBagLayout2);
        this.getContentPane().add(Panel2,
            new GridBagConstraints(0, 1, 1, 1, 1.0, 1.0,
            GridBagConstraints.CENTER, GridBagConstraints.HORIZONTAL,
                new Insets(1, 0, 1, 4), 190, 15));

        this.getContentPane().add(Panel1,
            new GridBagConstraints(0, 0, 1, 1, 1.0, 1.0,
```

```
            GridBagConstraints.CENTER, GridBagConstraints.HORIZONTAL,
                new Insets(0, 0, 1, 4), 23, 0));

        this.setResizable(false);
        this.setTitle("Calculator1");
        this.addWindowListener(new java.awt.event.WindowAdapter()
        {   public void windowClosing(WindowEvent e)
            {   System.exit(0);
            }
        });
    }

    public static void main(String[] args)
    {   Calculator1 mf=new Calculator1();
        mf.setBounds(300, 200, 240, 172);
        mf.show();
    }
}
```

9.4 编写一个程序,使之具有如主教材图 9-29 所示的界面,并且实现一些自己设计的功能。

解:本例中,目标窗口中部使用选择框显示图形选项,左侧为颜色选项。本例实现的功能是在画布上画图。在选择框中选中一种图形后,将在画布上画出这种图形。例如,图 9-4 中,当选择 Circle 时,将画出一个圆。该窗口中,还有复选框 filled,表示所画图形是否是实心图。程序对该复选框也做出响应。当选中该复选框时,所画的图形是填充了颜色的图形;不选中时,只画轮廓。

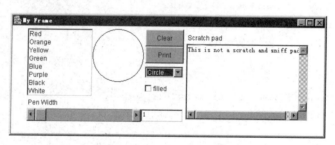

图 9-4 习题 9.4 的目标窗口

在此基础上,读者可以进一步增加功能。例如,可以使用左侧列表中的颜色画图,这部分功能留给读者完成。

Java 提供的作图方法中没有直接画三角形的方法,它提供的是一个应用面更广的方法,即 drawPolygon(int[] xPoints, int[] yPoints, int nPoints)。该方法可画含 nPoints 个点的多边形,nPoints 个点的横、纵坐标分别放在两个数组 xPoints 和 yPoints 中。程序中令 nPoints=3,并定义了 3 个点的坐标分别为(0,0)、(50,80)和(80,50)。对应于 drawPolygon()方法的是 fillPolygon()方法,后者将填充一个多边形。

程序代码实现如下:

```java
import java.awt.*;
import java.awt.event.*;
import java.util.*;
import javax.swing.*;
import javax.swing.event.*;

public class DrawGraphics extends WindowAdapter
    implements ActionListener, ItemListener, KeyListener, MouseListener
{   //定义窗口中必要的元素
    JFrame f;
    JButton clearJButton, printJButton;
    JLabel scratchPadJLabel,penWidthJLabel;
    JTextArea scratchPad;
    JComboBox shapeJComboBox;
    JList colorJList;
    JPanel jdraw;
    JTextArea ct;
    JCheckBox filledCheckbox;
    JPanel leftJPanel, rightJPanel, JButtonJPanel, penWidthJPanel;
    JScrollBar penWidthSlider;
    JTextField penWidthDisplay;
    int shapeTag=0;
    int colorTag=0;
    int filledTag =0;
    Color MyColor;

    class MyJPanel extends JPanel                  //显示画图的结果
    {   public void paint(Graphics g)
        {   int nPoints=3;
            int [] xPoints={0, 50, 80};
            int [] yPoints={0, 80, 50};

            if (filledTag==1)                       //填充
                switch(shapeTag)
                {   case 0:   g.fillRect(0, 0, 96, 96);
                        break;
                    case 1:   g.fillOval(0, 0, 96, 96);
                        break;
                    case 2:   g.fillPolygon(xPoints, yPoints, nPoints);
                        break;
                } //end of switch(shapeTag)
            else                                    //不填充,直接绘图
```

```java
                switch(shapeTag)
                {   case 0:     g.drawRect(0, 0, 96, 96);
                            break;
                    case 1:     g.drawOval(0, 0, 96, 96);
                            break;
                    case 2:     g.drawPolygon(xPoints, yPoints, nPoints);
                            break;
                } //end of switch(shapeTag)
        } //end of paint
    } //end of MyJPanel
    public static void main(String args [])
    {   DrawGraphics be=new DrawGraphics();
        be.go();
    }

    public void go()
    {   f=new JFrame("My JFrame");
        f.getContentPane().setLayout(new BorderLayout());
        f.setSize(600, 300);

        leftJPanel=new JPanel();
        leftJPanel.setLayout(new BorderLayout());
        rightJPanel=new JPanel();
        rightJPanel.setLayout(new BorderLayout());
        JButtonJPanel=new JPanel();
        JButtonJPanel.setLayout(new GridLayout(4,1));
        penWidthJPanel=new JPanel();
        penWidthJPanel.setLayout(new BorderLayout());
        Vector colors=new Vector();                     //定义颜色向量
        colors.addElement("Red");
        colors.addElement("Orange");
        colors.addElement("Yellow ");
        colors.addElement("Green");
        colors.addElement("Blue");
        colors.addElement("Purple");
        colors.addElement("Black");
        colors.addElement("White");

        colorJList=new JList(colors);                   //加入列表中

        JScrollPane jsp=new JScrollPane(colorJList,
            JScrollPane.VERTICAL_SCROLLBAR_AS_NEEDED,
            JScrollPane.HORIZONTAL_SCROLLBAR_AS_NEEDED);
```

```java
leftJPanel.add(jsp,BorderLayout.WEST);

jdraw=new MyJPanel();
jdraw.setSize(200,200);
leftJPanel.add(jdraw,BorderLayout.CENTER);

clearJButton=new JButton("Clear");
clearJButton.addActionListener(this);

printJButton=new JButton("Print");
shapeJComboBox=new JComboBox();                //定义选择图案的下拉菜单
shapeJComboBox.addItem("Square");
shapeJComboBox.addItem("Circle");
shapeJComboBox.addItem("Triangle");
shapeJComboBox.addItemListener(this);
filledCheckbox=new JCheckBox("filled", true);
filledCheckbox.addItemListener(new ItemListener()
{   public void itemStateChanged(ItemEvent e)
    {   if (filledTag==1) filledTag=0;
        else filledTag=1;
    }
});

JButtonJPanel.add(clearJButton);
JButtonJPanel.add(printJButton);
JButtonJPanel.add(shapeJComboBox);
JButtonJPanel.add(filledCheckbox);

leftJPanel.add(JButtonJPanel,BorderLayout.EAST);

penWidthJLabel=new JLabel("Pen Width");
penWidthSlider=new JScrollBar(Scrollbar.HORIZONTAL, 1, 1, 1, 10);
penWidthDisplay=new JTextField("1", 8);

penWidthJPanel.add(penWidthJLabel,BorderLayout.NORTH);
penWidthJPanel.add(penWidthSlider,BorderLayout.CENTER);
penWidthJPanel.add(penWidthDisplay,BorderLayout.EAST);

leftJPanel.add(penWidthJPanel,BorderLayout.SOUTH);

scratchPadJLabel=new JLabel("Scratch pad");
rightJPanel.add(scratchPadJLabel,BorderLayout.NORTH);
```

```
        scratchPad=new JTextArea("This is not a scratch and sniff pad.", 8,
20);
        JScrollPane jsp2=new JScrollPane(scratchPad,
            JScrollPane.VERTICAL_SCROLLBAR_AS_NEEDED,
            JScrollPane.HORIZONTAL_SCROLLBAR_AS_NEEDED);

        rightJPanel.add(jsp2,BorderLayout.CENTER);

        f.add(leftJPanel,BorderLayout.CENTER);
        f.add(rightJPanel,BorderLayout.EAST);

        f.addWindowListener(this);
        f.setVisible(true);
    }

    //实现 ActionJListener 接口中的 actionPerformed()方法
    public void actionPerformed(ActionEvent e)
    {   //clearJButton
        if (e.getSource()==clearJButton)
        {   System.exit(0);
        }
    }

    //实现 ItemJListener 接口中的 itemStateChanged()方法
    public void itemStateChanged(ItemEvent e)
    {   if (e.getSource() instanceof JComboBox)
        {   shapeTag=((JComboBox)e.getSource()).getSelectedIndex();
            jdraw.repaint();
        }
    }


    public void mouseMoved (MouseEvent e) {}

    //实现 MouseJListener 接口中的方法
    public void mouseClicked (MouseEvent e) {}

    public void mouseEntered (MouseEvent e) {}
    public void mouseExited (MouseEvent e) {}
    public void mousePressed (MouseEvent e) {}
    public void mouseReleased (MouseEvent e) {}

    public void keyTyped(KeyEvent evt){}
```

```
    public void keyPressed(KeyEvent evt){}
    public void keyReleased(KeyEvent evt){}

    public void windowClosing(WindowEvent e)
    {   System.exit(0);
    }
}
```

第 10 章 Java Applet

10.1 什么是 Applet？它与普通应用程序有什么区别？

解：Applet 也叫小应用程序，是在浏览器环境下运行的一类特殊的 Java 程序。我们知道，Java 语言作为一种解释型语言，它的字节码程序需要一个专门的解释器来执行它。对于应用程序来说，这个解释器是独立的软件，如 JDK 的 java.exe。字节码也可能通过 Web 来传输，然后由 Web 浏览器中所含的 Java 解释器来执行。这就是 Java Applet，对应的解释器就是 Internet 的浏览器软件。

Applet 的基本工作原理也与应用程序有所不同。一般来说，Applet 需要和 HTML 文件配合使用。HTML 页必须告诉浏览器装载哪个 Applet，Applet 应该装载在什么位置。

首先，Java 将编译好的字节码文件也就是类文件保存在特定的 WWW 服务器上，同一个或另一个 WWW 服务器上保存着嵌入了该字节码文件名的 HTML 文件。当某一个浏览器向服务器请求下载嵌入了 Applet 的 HTML 文件时，该文件就从 WWW 服务器上下载到客户端，由 WWW 浏览器解释 HTML 中的各种标记，按照其约定将文件中的信息以一定的格式显示在用户屏幕上。

Applet 与普通应用程序的不同可分为两方面。

(1) Applet 的运行

由于 Applet 是在浏览器中运行的，因此它不能通过直接键入一条命令来启动。在运行 Applet 时，必须首先创建一个 HTML 文件，在该文件中通过 <applet> 标记指定要运行的 Applet 程序名，然后将该 HTML 文件的 URL 通知浏览器，最后通过浏览器装入并运行该 Applet 程序。Applet 不需要建立自己的主流程框架，它实现的功能也是不完全的。它需要和浏览器中预先实现好的功能结合起来。

普通应用程序总是从 main() 方法开始执行，然而 Applet 与此不同，Applet 程序是从构造方法开始执行的。在构造方法执行结束以后，浏览器调用 Applet 的 init() 方法，该方法完成 Applet 的初始化操作。在 init() 方法执行结束以后，浏览器又调用一个名为 start() 的方法。

不论是 init() 方法还是 start() 方法，它们都是在 Applet 被激活前执行的，因此不能用它们来实现 Applet 的功能。事实上，与一般应用程序中的 main() 方法不同，在 Applet 中，没有任何一个方法在程序的整个生命周期内自始至终一直运行。

(2) Applet 的安全性限定

Applet 是可以通过网络传输和装载的程序，为了避免网络上不安全事件的发生，Java 提供了一个 SecurityManager 类，该类在 Java 虚拟机(JVM)上对几乎所有系统级调用进行监控。这个工作模式称为 sandbox（沙箱）安全模式——JVM 提供一个沙箱，允许

Applet 在其中运行。一旦 Applet 企图离开沙箱,它的运行就会被禁止。

在后来的 JDK 1.1 版本中,提出了"签名 Applet"的概念。有正确签名的 Applet 与本地代码一样,可以使用本地的资源。没有签名的 Applet 还与前一版本一样,只在沙箱中运行。

在 Java 2 平台下,安全机制又有较大改善。它允许用户自己设定相关的安全级别。

对系统安全性的限定尺度通常是在浏览器中设定的。几乎所有浏览器都禁止 Applet 程序的下述行为:

- 运行过程中调用执行另一个程序。
- 所有的文件 I/O 操作。
- 调用本机(native)方法。
- 企图打开提供该 Applet 的主机以外的某个套接口(socket)。

10.2 简述 paint()、repaint()和 update()三个方法之间的调用关系。

解:paint()方法的主要作用是在 Applet 的界面中绘制文字、图形等要显示的内容。Applet 被启动之后,系统自动调用 paint()来绘制界面。在浏览器中,每当 Applet 显示内容需要刷新时,都会调用 paint()方法。Applet 的其他相关方法被调用时,系统也会相应地调用 paint()方法。

当程序重新更新显示内容时,使用 repaint()方法通知系统去更新显示内容。此时 AWT 线程会自动调用 update()方法,该方法首先将当前显示画面清空,然后调用 paint()方法绘制新的显示内容。

这三者之间的调用关系如图 10-1 所示。

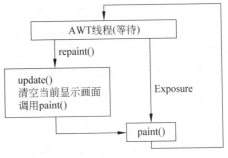

图 10-1 paint()、repaint()和 update() 之间的调用关系

10.3 编写一个 Applet 在屏幕中画一组同心圆,相邻两个圆直径大小相差 10(pixel),Applet 的大小为 300×300(pixel)。

解:使用 Applet 画图,要建立自己定义的 Applet 类的子类,同时还要建立一个对应的 HTML 文件。

本例中,自己定义的子类是 DrawOval。它首先使用 g.getClipBounds().getSize()得到在 Web 页面中 Applet 所占的空间大小,然后在这个范围内连续画圆。

Java 中提供的画圆方法是:

```
public abstract void drawOval(int x, int y, int width, int height)
```

实际上这是一个画椭圆的方法。圆是椭圆的一个特例。当椭圆的长轴与短轴相等时,就变为圆。使用 drawOval()画圆时,最后的两个参数(椭圆所占的矩形框的宽和高)取等值就可以画圆了。

drawOval()方法的前两个参数 x 和 y 是矩形框的左上角的横、纵坐标值。这里已知 cx 和 cy 分别是矩形框的中心点坐标,radius 是圆的半径,通过计算 cx-radius 和 cy-radius 可以得到矩形框左上角的坐标值。因为是画一组同心圆,每次要增大圆的半径,因此左上角也需要移动。使用 num 记录所画的圆的个数,同时可由它计算出下一个圆外切矩形的左上角。

程序代码实现如下:

```
import java.awt.*;
import java.applet.*;

public class DrawOval extends Applet              //继承于 Applet
{   public void paint(Graphics g)
    {   int w, h, cx, cy, mradius, radius=5;
        int num=1;
        int l=radius*2;                            //初始时,画直径为10的圆
        Dimension dm=g.getClipBounds().getSize();  //得到画图的矩形尺寸

        w=dm.width;                                //矩形的宽
        h=dm.height;                               //矩形的长
        cx=w/2;                                    //圆心的横坐标
        cy=h/2;                                    //圆心的纵坐标
        mradius=Math.min(w,h);                     //取最小值,得到正方形

        while (l <=mradius)                        //不出正方形边界时
        {   g.drawOval(cx - radius * num, cy-radius * num, l, l);
            l=l+2 * radius;                        //直径增大10
            num=num+1;
        }
    }
    public void init()
    {   resize(200, 150);
    }
}
```

对应的最简单的 HTML 文档如下:

```
<html>
<applet code=DrawOval.class width=300 height=300>
</applet>
</html>
```

运行该 HTML 程序,在浏览器中将画出一组同心圆,执行的结果如图 10-2 所示。

10.4 编写一个 Applet 在屏幕上画椭圆,椭圆的大小和位置由鼠标决定(在左上角按下鼠标,拖动到右下角放开鼠标,在由此决定的矩形中画椭圆)。

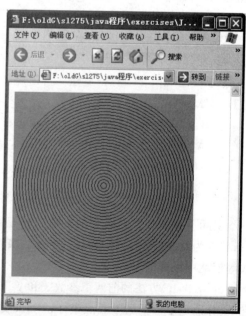

图 10-2 同心圆示例

解：在 Applet 中，通过鼠标确定一个矩形框，并在框内画椭圆。为能获取鼠标单击点的位置信息，需要用鼠标事件监听程序 MouseListener 和 MouseMotionListener 来获取鼠标事件，为此说明一个类 DrawOval2，来实现这两个接口。DrawOval2 已继承于 Applet，因为 Java 中没有多重继承机制，所以它不能再继承于 MouseAdapter 类，而只能实现监听程序接口。程序中需要实现这两个接口中的全部方法，包括：mouseClicked()、mouseEntered()、mouseExited()、mousePressed()、mouseReleased()、mouseMoved()和 mouseDragged()。对于用不到的方法，可以令方法体为空。

由于 Applet 也是一个组件，继承于 Component，同样需使用 addMouseListener(this)和 addMouseMotionListener(this)来添加事件监听程序，并在类中实现鼠标按下(mousePressed)和拖拉(mouseDragged)方法。鼠标按下时得到画圆区域的左上角位置(leftPx, leftPy)，拖动过程中得到这个区域的右下角位置(rightPx, rightPy)。因为是在鼠标拖动后得到右下角坐标，所以要重新实现 public void mouseDragged(MouseEvent e) 方法。

程序代码实现如下：

```
import java.awt.*;
import java.awt.event.*;
import java.applet.*;
public class DrawOval2 extends Applet implements MouseListener, MouseMotionListener
{   int rightPx, rightPy;
    int leftPx, leftPy;

    public void paint(Graphics g)
    {   g.drawOval(leftPx, leftPy, rightPx-leftPx, rightPy-leftPy);
    }

    public void init()
    {   resize(300, 250);
        addMouseListener(this);
        addMouseMotionListener(this);
    }

    public void mouseClicked(MouseEvent e) {}
    public void mouseMoved(MouseEvent e) {}
    public void mouseDragged(MouseEvent e)
    {   rightPx=e.getX();                    //拖拉鼠标时的坐标
        rightPy=e.getY();
        repaint();
    }
    public void mouseEntered (MouseEvent e){}
    public void mouseExited (MouseEvent e){}
```

```
        public void mousePressed (MouseEvent e)
        {   leftPx=e.getX();                         //按下鼠标时的坐标
            leftPy=e.getY();
        }
        public void mouseReleased (MouseEvent e) {}
}
```

对应的 HTML 文档如下：

```
<html>
<applet code=DrawOval2.class width=300 height=300>
</applet>
</html>
```

程序的执行结果如图 10-3 所示。

10.5 修改习题 10.3 中的 Applet，使它能够在鼠标移动的过程中动态地改变同心圆的大小。

解：结合前面的习题 10.3、10.4，在鼠标按下及拖动的范围内画同心圆。鼠标按下时得到画圆区域的左上角位置(leftPx，leftPy)，拖动过程中得到这个区域的右下角位置(rightPx，rightPy)。为了得到整个的同心圆，使用下面的语句：

图 10-3 画椭圆示例

```
rightPx=Math.min(dm.width, rightPx);
rightPy=Math.min(dm.height, rightPy);
```

来控制右下角位置不要越出 Applet 的区域。

读者可以注释掉这两个语句，看看执行效果是什么样的。

程序运行时，随着鼠标指针的移动，可以得到大小及位置不同的一组同心圆。与前一例类似，因为是在鼠标拖动后得到右下角坐标，所以要重新实现方法 public void mouseDragged(MouseEvent e)。根据得到的坐标值，计算出圆的外切正方形边界，画圆时不能超出边界，使用 mradius＝Math. min(w，h)；语句得到最大半径。

程序代码实现如下：

```
import java.awt.*;
import java.applet.*;
import java.awt.event.*;

public class DrawOval5 extends Applet implements MouseListener, MouseMotionListener
{   int rightPx, rightPy;
    int leftPx, leftPy;

    public void paint(Graphics g)
    {   int w, h, cx, cy, mradius, radius=5;
        int num=1;
```

```java
        int l=radius*2;
        Dimension dm=g.getClipBounds().getSize();    //得到画图的矩形尺寸

        rightPx=Math.min(dm.width, rightPx);
        rightPy=Math.min(dm.height, rightPy);
        w=rightPx-leftPx;
        h=rightPy-leftPy;
        cx=(leftPx+rightPx)/2;
        cy=(leftPy+rightPy)/2;
        mradius=Math.min(w, h);

        while (l<=mradius)
        {   g.drawOval(cx-radius*num, cy-radius*num, l, l);
            l=l+2*radius;
            num=num+1;
        }
    }

    public void init()
    {
        resize(300, 250);
        addMouseListener(this);
        addMouseMotionListener(this);
    }

    public void mouseClicked(MouseEvent e) {}

    public void mouseMoved(MouseEvent e) {}
    public void mouseDragged(MouseEvent e)
    {
        rightPx=e.getX();
        rightPy=e.getY();
        repaint();
    }
    public void mouseEntered (MouseEvent e){}
    public void mouseExited (MouseEvent e){}
    public void mousePressed (MouseEvent e)
    {
        leftPx=e.getX();
        leftPy=e.getY();
    }
    public void mouseReleased (MouseEvent e) {}
}
```

对应的 HTML 文档如下：

```
<html>
<applet code=DrawOval5.class width=300 height=300>
</applet>
</html>
```

程序执行时,使用鼠标先选择起始点,然后拖动到另一个点,在这两个点决定的矩形框内画同心圆。运行结果如图 10-4 和图 10-5 所示。

图 10-4　鼠标指针在第一个位置

图 10-5　鼠标指针继续向右下方拖动后

10.6　编写一个 Applet,在随机的位置上画出几个随机大小的矩形。如果一个矩形的宽度小于高度,则矩形填充成亮紫色;如果矩形的宽度大于高度,则矩形填充为浅黄色;如果矩形的宽度与高度相等,则只用红色线画出矩形的边框。

解：使用 Random 类获取矩形个数、矩形起始点坐标和矩形宽度及高度。应用 DrawRect(x, y, width, height)方法画矩形,其中 x、y 是矩形左上角的坐标值,width 和 height 是矩形的宽和高。

程序代码实现如下：

```java
import java.applet.Applet;
import java.awt.*;
import java.util.*;

public class DrawRect1 extends Applet
{   public DrawRect1()
    {   super();                                    //调用父类的构造函数
    }

    public void paint(Graphics g)
    {   Random random=new Random();

        int maxNumber=random.nextInt(15);           //矩形个数
        int maxWidth=g.getClipBounds().width;       //最大宽度
        int maxHeight=g.getClipBounds().height;     //最大高度
```

```java
        /***画 maxNumber 个矩形***/
        for(int i=0;i<maxNumber;i++){
            int startPx=random.nextInt(maxWidth);            //起始点横坐标
            int startPy=random.nextInt(maxHeight);           //起始点纵坐标
            int width=random.nextInt(maxWidth-startPx);      //矩形的宽
            int height=random.nextInt(maxHeight-startPy);    //矩形的高

            /***画矩形,根据宽高关系填充颜色***/
            if(width<height){                                //宽小于高
                g.setColor(Color.MAGENTA);
                g.fillRect(startPx, startPy, width, height); //实心图
            }
            else if(width>height)                            //宽大于高
            {   g.setColor(Color.YELLOW);
                g.fillRect(startPx, startPy, width, height); //实心图
            }
            else                                             //正方形
            {   g.setColor(Color.RED);
                g.drawRect(startPx, startPy, width, height); //只画边框
            }
        }
    }

    public void init()
    {   this.setSize(300, 300);
    }
}
```

对应的 HTML 文档如下:

```
<html>
< applet code = DrawRect1. class width = 300 height=300>
</applet>
</html>
```

程序的执行结果如图 10-6 所示。

10.7 编写一个 Applet,以等间距画出等宽的几个竖条,竖条的高度随机。将最高的竖条填充为黄色,最矮的竖条填充为绿色,其余的竖条填充为蓝色。

解:竖条即是矩形。应用方法 DrawRect(int x, int y, int width, int height)完成画竖条的工作。本例中要求以等间距来画竖条,所以矩形起始点横坐标

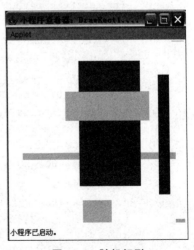

图 10-6 随机矩形

的获取是关键点。采用的方法是在 0~maxWidth 区间上生成一个等差数列,其中 maxWidth 表示 applet 的宽度。

数组可用来保存相同类型的多个对象,但因为竖条的个数不定,所以本例中使用了可变长数组 ArrayList 类型。这是 List 接口的一个可变长数组,它实现了所有 List 接口的操作,并允许存储 null 值。ArrayList 使用数组存储元素,在其自动变长机制中,应用 ensureCapacity()方法来保证数组长度,即当数组长度不足以容纳元素时,将数组容量扩充为原长度的 1.5 倍。定义的 ArrayList 类型的可变长数组 arrayList 用来保存 rectNumber 个起始点的横坐标。当竖条的个数增多而 arrayList 数组不能容纳时,系统自动增大 arrayList 的大小,以满足需要。

因为最高的竖条和最矮的竖条需要填充与其他竖条不同的颜色,所以需要先把竖条都创建完毕,以决定哪个是最高的,哪个是最矮的,而不能边画边生成。

程序代码实现如下:

```java
import java.applet.Applet;
import java.awt.*;
import java.util.*;

public class DrawRect2 extends Applet
{   public DrawRect2()
    {   super();
    }

    public void paint(Graphics g)
    {   Random random=new Random();
        int maxWidth=g.getClipBounds().width;
        int maxHeight=g.getClipBounds().height;

        int dist=random.nextInt(maxWidth/9);                    //间距
        int rectNumber=maxWidth/dist;

        /***获取 rectNumber 个起始点横坐标,存入 arrayList 中***/
        ArrayList<Integer>arrayList=new ArrayList<Integer>();
        int firstItem=maxWidth-rectNumber*dist;
        for(int i=0; i<rectNumber; i++)
        {   arrayList.add(firstItem+i*dist);
        }

        int[] startPy=new int[rectNumber];                      //纵坐标
        for(int i=0; i<rectNumber; i++)
        {   while(startPy[i]==0)
                startPy[i]=random.nextInt(maxHeight);
        }
```

```
        int min=startPy[0];
        int max=startPy[0];
        for(int i=1;i<rectNumber;i++)
        {   if(startPy[i]<min) min=startPy[i];                    //取最大值
            if(startPy[i]>max) max=startPy[i];                    //取最小值
        }

        int width=random.nextInt(dist);                           //获取宽度值
        for(Integer startPx : arrayList)
        {   int height=startPy[arrayList.indexOf(startPx)];       //获取高度值
            if(height==min)
            {   g.setColor(Color.GREEN);
                g.fillRect(startPx, maxHeight-height, width, height);
            }
            else if(height==max)
            {   g.setColor(Color.YELLOW);
                g.fillRect(startPx, maxHeight-height, width, height);
            }
            else
            {   g.setColor(Color.BLUE);
                g.fillRect(startPx, maxHeight-height, width, height);
            }
        }
    }

    public void init()
    {   this.setSize(300, 300);
    }
}
```

对应的 HTML 文档如下：

```
<html>
< applet code = DrawRect2. class width = 300 height=300>
</applet>
</html>
```

程序的执行结果如图 10-7 所示。

10.8 编写一个 Applet,画出 20 条水平的、随机颜色的平行线。要求线的长度相同,整条线段都要在可视区域内。

解：可以调用 DrawLine(int x1, int y1, int x2, int y2)方法完成画线工作。本例的重点在于线段端点纵坐标的获取。因为是水平方向的平行线,所以

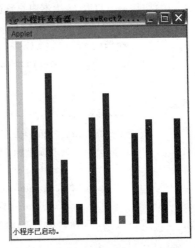

图 10-7　等间距随机长度的竖条

各条线的端点纵坐标不能相同,本例中使用 Set 类来保证这一点。

  Set 类与 List 一样同属于容器类 Collections,Set 类要求其内部元素具有唯一性,这样就可以保证没有重复的值了。有三种类型的 Set,即 HashSet、TreeSet 和 LinkedHashSet。其中,HashSet 采用散列函数对元素进行计算,根据得到的值安排其存储位置,便于进行快速查找。存入 HashSet 的对象必须定义 hashCode。TreeSet 是具有特定次序的集合,底层为树型结构,应用 TreeSet 可以获取有序的序列,其元素排序采用红黑树机制。LinkedHashSet 则是链式散列集合,在内部使用散列以加快查询速度,同时使用链表维护元素次序,使元素按照插入顺序排列。

  本例中还要求所画的整条线段都要在可视区域内,可以使用如下方法来实现:首先取可视区域的宽度 width,然后取小于 width 的随机数作为线段长度 length,并取小于(width-length)的随机数作为线段起始点横坐标。

  程序代码实现如下:

```java
import java.applet.Applet;
import java.awt.*;
import java.util.*;

public class DrawLine1 extends Applet
{   public DrawLine1()
    {   super();
    }

    public void paint(Graphics g)
    {   Random random=new Random();

        int width=g.getClipBounds().width;              //宽度
        int length=random.nextInt(width);               //线的长度,要小于 width

        /***获取 20 个不同起始点纵坐标***/
        Set<Integer>set=new HashSet<Integer>();
        while(set.size()<20)
        {   set.add(random.nextInt(g.getClipBounds().height));
        }

        /***画线***/
        for(Integer startPy : set)
        {       //获取横坐标,要小于(width-length),保证不出界
            int startPx=random.nextInt(width-length);
            g.setColor(new Color(random.nextInt(256), random.nextInt(256),
                random.nextInt(256)));                              //设置颜色
            g.drawLine(startPx, startPy, startPx+length, startPy);  //画水平线
        }
```

	}
	public void init()
	{ this.setSize(300, 200);
	}
```

对应的 HTML 文档如下：

```
<html>
<applet code=DrawLine1.class width=300 height=300>
</applet>
</html>
```

程序的执行结果如图 10-8 所示。

10.9 编写一个 Applet，画出 20 条垂直方向的、随机长度的平行线。要求线的颜色随机，整条线段都要在可视区域内。

解：实现过程与习题 10.8 类似。不同之处在于此题目中需要获取 20 个不同的横坐标值，另外需要随机获取 20 个值作为线的长度。

图 10-8　水平方向等长平行线

为保证整条线段都在可视区内，用如下方法实现：首先取可视区域高度 maxLength，然后取小于 maxLength 的随机数作为线段起始点纵坐标 startPy，取小于（maxLength-startPy）的随机数作为线段长度。

程序代码实现如下：

```java
import java.applet.Applet;
import java.awt.*;
import java.util.*;

public class DrawLine2 extends Applet
{   public DrawLine2()
	{   super();
	}

	public void paint(Graphics g)
	{   Random random=new Random();

		int maxLength=random.nextInt(g.getClipBounds().height);    //最大长度

		/***获取 20 个不同起始点的横坐标***/
		Set<Integer>set=new HashSet<Integer>();
		while(set.size()<20)
		{   set.add(random.nextInt(g.getClipBounds().width));
```
```

```
        }

    /***画20条垂直方向的平行线***/
    for(Integer startPx : set)
    {   //获取起始点纵坐标,要小于maxLength
        int startPy=random.nextInt(maxLength);
        //获取线的长度,要小于(maxLength - startPy)
        int length=random.nextInt(maxLength-startPy);
        g.setColor(new Color(random.nextInt(256), random.nextInt(256),
            random.nextInt(256)));                                    //设置颜色
        g.drawLine(startPx, startPy, startPx, startPy+length);  //画垂直线

    }
    }

    public void init()
    {   this.setSize(300, 300);
    }
```

对应的HTML文档如下：

```
<html>
<applet code=DrawLine2.class width=300 height=300>
</applet>
</html>
```

程序的执行结果如图10-9所示。

10.10 编写一个Applet,随机选择矩形、圆形、椭圆、直线等形状,在可视区域内绘制20个图形,同一种图形使用同一种颜色,不需要填充。

解：定义一个ArrayList用于存放图形类型,元素个数为20。元素值在0、1、2、3中随机选取,其中0代表矩形,1代表圆形,2代表椭圆,3代表直线。定义一个数组shapeColor,存放4种随机生成的颜色。遍历ArrayList,根据其元素值画不同类型的图形。对同一种图形,在shapeColor中选择同一种颜色。

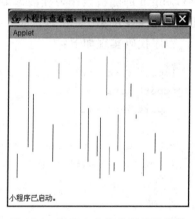

图10-9 垂直方向长度随机平行线

程序代码实现如下：

```
import java.applet.Applet;
import java.awt.*;
import java.util.*;

public class DrawShape extends Applet
{   public DrawShape()
```

```java
{   super();
}

public void paint(Graphics g)
{   Random random=new Random();

    int shapeNumber=20;
    int width=g.getClipBounds().width;
    int height=g.getClipBounds().height;

    int[] startPx=new int[20];
    int[] startPy=new int[20];

    //**图形类型,0代表矩形,1代表圆形,2代表椭圆,3代表直线
    //**存入 ArrayList 中**
    ArrayList<Integer>arrayList=new ArrayList<Integer>();
    while(arrayList.size()<shapeNumber)
    {   arrayList.add(random.nextInt(4));    //随机得到0~3的整数,代表4种图形
    }

    /***随机选择4种图形颜色,存入数组 shapeColor 中***/
    Color[] shapeColor=new Color[4];
    for(int i=0;i<4;i++)
    {   shapeColor[i]=new Color(random.nextInt(256), random.nextInt(256),
            random.nextInt(256));
    }

    /***画20个随机图形***/
    for(Integer i :arrayList)
    {   g.setColor(shapeColor[i]);
        startPx[i]=random.nextInt(width);
        startPy[i]=random.nextInt(height);
        switch(i)                            //根据已得到的图形分类,分别画出图形
        {   case 0:                          //画矩形
                g.drawRect(startPx[i], startPy[i],
                    random.nextInt(width-startPx[i]),
                    random.nextInt(height-startPy[i]));
                break;
            case 1:                          //画圆
                int diameter=Math.min(random.nextInt(width-startPx[i]),
                    random.nextInt(height-startPy[i]));
                g.drawOval(startPx[i], startPy[i], diameter, diameter);
                break;
            case 2:                          //画椭圆
```

```
                g.drawOval(startPx[i], startPy[i],
                    random.nextInt(width-startPx[i]),
                    random.nextInt(height-startPy[i]));
                break;
            case 3:                                            //画直线
                g.drawLine(startPx[i], startPy[i],
                    random.nextInt(width),
                    random.nextInt(height));
                break;
            default:
                break;
        } //end of switch(i)
    }//end of for loop
}//end of paint

public void init()
{   this.setSize(300, 300);
    }
}
```

对应的 HTML 文档如下：

```
<html>
<applet code=DrawShape.class width=300 height=300>
</applet>
</html>
```

程序的执行结果如图 10-10 所示。

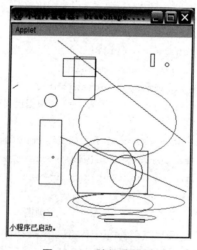

图 10-10　随机画图形

第 11 章 Java 数据流

11.1 列举 Java 所包括的输入/输出操作,完成所有输入/输出操作所需的类都包含于哪个软件包中?

解:输入/输出操作是指与外部设备或其他计算机进行信息交互的操作。例如从键盘读取数据,从文件中获取或者向文件写入数据,在显示屏上显示数据,以及通过网络与其他计算机进行信息交互等。

Java 的输入/输出操作必须借助于输入/输出类库来完成,这个类库包含于 java.io 包中。这个包中的类大部分是用来完成流式输入/输出操作的。

11.2 什么叫作流?输入/输出流分别对应哪两个抽象类?

解:流也称数据流,是指在计算机的输入/输出之间流动的数据的序列。这是一组有顺序的、有起点和终点的字节集合。一般分为两种流:可从中读出一系列字节的对象称为输入流,能向其中写入一系列字节的对象称为输出流。

输入流用抽象类 InputStream 来实现,输出流用抽象类 OutputStream 来实现。

11.3 InputStream 有哪些直接子类?其功能是什么?

解:Java JDK 8 中提供的 InputStream 流的主要直接子类列在表 11-1 中。

表 11-1 InputStream 流的主要直接子类

子 类 名	功　　能
AudioInputStream 音频输入流	可以输入特定格式和长度的音频对象
ByteArrayInputStream 字节数组输入流	含有一个内部的缓冲区,要读入的数据先进入缓冲区,然后真正的读入操作实际上是对该缓冲区进行的。使用缓冲区可以有效地提高输入/输出效率
FileInputStream 文件输入流	用来进行文件读入操作,可以从文件系统的一个文件中获取输入字节。至于可操作的文件类型,则要依赖于主机环境。如果所指定的文件不存在,则产生 FileNotFoundException 异常
FilterInputStream 过滤输入流	在输入数据的同时能对所传输的数据做指定类型或格式的转换,它包含其他的输入流作为基本的数据源
ObjectInputStream 对象流	可以将对象数据写入文件数据流中,因此数据流的基本操作单位不仅仅是基本数据类型,还可以是对象
PipedInputStream 管道输入流	主要用于线程间的通信。管道输入流应该与管道输出流一起使用,建立一个通信通道。一个线程从管道输入流对象读入数据,再由另外的线程将数据写到对应的管道输出流中
SequenceInputStream 顺序输入流	可以把多个其他的输入流逻辑连接起来,形成一个完整的输入流。其他的输入流表示成一个有序序列,顺序输入流从第一个输入流开始读入数据,直到结尾;然后读入第二个输入流,依此类推,按照有序序列中的输入流顺序,一个一个地读入,直到读入完最后一个输入流

11.4 OutputStream 有哪些直接子类？其功能是什么？

解：Java JDK 8 中提供的 OutputStream 流的主要直接子类列在表 11-2 中。

表 11-2 OutputStream 流的主要直接子类

子 类 名	功 能
ByteArrayOutputStream 字节数组输出流	与字节数组输入流的数据流向相反，输出流中的数据先写入字节数组缓冲区中
FileOutputStream 文件输出流	用来完成文件输出操作
FilterOutputStream 过滤输出流	它是所有过滤输出流类的超类。它以其他的输出流作为基本的数据源，在输出数据的同时能对所传输的数据做指定类型或格式的转换
ObjectOutputStream 对象输出流	它可以将基本数据类型和图形对象写到输出流中
PipedOutputStream 管道输出流	它用来与管道输入流相连，形成一个通信管道

11.5 实现一个输入程序，接收从键盘读入的字符串。当字符串中所含字符个数少于程序设定的上限时，输出这个字符串；否则抛出 MyStringException1 异常，在异常处理中要求重新输入新的字符串或中断程序运行。

解：根据题意，需要接收从键盘读入的字符串，为此先定义相应的对象，使用的方法如下：

```
BufferedReader keyboard=new BufferedReader(new InputStreamReader(System.in));
```

然后接收读入的一行内容，方法如下：

```
String input=keyboard.readLine();
```

使用 input.length() 方法可以得到输入字符串的长度，并判断是否符合条件。程序中还定义了异常类 MyStringException1，以处理字符串中所含字符个数过多的情况。

程序代码实现如下：

```
import java.io.BufferedReader;
import java.io.IOException;
import java.io.InputStreamReader;

public class MyStringException1 extends Exception          //异常类的定义
{   public MyStringException1()
    {   super("字符串长度过大,重新输入字符串");
    }

    public MyStringException1(String message)
    {   super(message);
    }
}
```

```java
public class Input
{   public static void main(String[] args) throws IOException
    {   //设置字符串长度上限
        int maxLength=Integer.parseInt(args[0]);        //运行时设置输入的限度

        //应用BufferedReader类捕获输入字符串
        BufferedReader keyboard=new BufferedReader
            (new InputStreamReader(System.in));

        try
        {   String input=keyboard.readLine();

            //如果输入字符串长度小于长度上限时,输出字符串
            if(input.length()<maxLength)
            {   System.out.println(input);
            }
            //捕获异常MyStringException1
            else
            {   throw new MyStringException1();
            }
        }catch(MyStringException1 e)
        {   System.out.println(e.getMessage());
        }
    }
}
```

11.6 利用输入/输出流编写一个程序,实现文件复制的功能。程序的命令行参数的形式及操作功能均类似于DOS中的copy命令。

解：DOS下的copy命令可以带较多的参数,这里我们定义一种较简单的形式,完成最基本的文件复制功能。实现的文件复制命令的格式如下：

copy source destination

其中,source是要被复制的文件名,destination是新生成的文件名称。由于两个文件都可以包含目录,所以需要在程序中对此进行处理。读者可以在上面命令的格式基础上,自行添加其他的参数,并实现之。

程序代码实现如下：

```java
import java.io.*;
import java.util.*;

public class copy
{   public static void main(String args[])
    {   errMes errorMes=new errMes();
```

```
            if(args.length<2)                        //需要有命令行参数
            {   errorMes.usage();
            }
            if(args.length==2)
            {   if(!new File(args[1]).isDirectory())  //判断目标文件名是否是文件
                {   copyfile cp=new copyfile();
                    cp.copy(args[0], args[1]);        //将命令行中的参数作为方法的参数
                    System.exit(0);
                }
                else                                  //目标文件是目录时的处理
                {   int index=args[0].lastIndexOf("\\");      //找到最后面的'\\'
                    //最后面\\之后的字符串应是文件名
                    String FileName=args[0].substring(index+1);
                    if(!args[1].endsWith("\\"))               //对目标文件名的处理
                        FileName=args[1]+"\\"+FileName;
                    else
                        FileName=args[1]+FileName;
                    copyfile cp=new copyfile();
                    cp.copy(args[0], FileName);
                    System.exit(0);
                }
            }

            Vector sourcename, sourceFiles;
            source sc=new source(args);
            sourcename=sc.getfiles();
            sourceFiles=sc.getSourceFiles();
            String sourcepath=sc.getpath();
            String des;
            destination desti=new destination(args);
            des=desti.getdirectory();
            for(int i=0; i<sourcename.size(); i++)
            {   String sf=(String)sourceFiles.get(i);
                String df=des+sourcename.get(i);
                copyfile cp=new copyfile();
                cp.copy(sf, df);
            }
        }
    }

class copyfile
{   public void copy(String from, String to)
    {   if(new File(to).exists())
        {   System.out.println("target file : " +
```

```java
                to+" exist. do you want to overwrite it? (y/n)" );
            int YorN=0;
            try
            {   BufferedReader in=new BufferedReader(
                    new InputStreamReader(System.in));
                YorN=in.read();
                if(YorN!='y' && YorN !='Y')
                    return;
            }catch(Exception e)
            {   e.printStackTrace();
            }
        }
        try
        {   FileInputStream fis=new FileInputStream(from);
            FileOutputStream fos=new FileOutputStream(to);
            int result=0;
            byte[] buf=new byte[100];
            while(result !=-1)
            {   result=fis.read(buf);
                if(result>0)
                    fos.write(buf, 0, result);
            }
            fis.close();
            fos.close();
            System.out.println("copy "+from+" to "+to);
        }catch(Exception  e)
        {   e.printStackTrace();
        }
    }
}

class source
{   Vector sourceFiles;
    Vector filenames;
    String filepath;
    String[] files;

    source(String fnames[])
    {   sourceFiles=new Vector();
        files=fnames;
        filepath=getfilepath(fnames[0]);
        for(int i=0; i<fnames.length-1; i++)
        {   if(new File(fnames[i]).isFile())
                sourceFiles.addElement(fnames[i]);
```

```
            }
        }

        Vector getSourceFiles()
        {   return sourceFiles;
        }

        Vector getfiles()
        {   Vector name=new Vector();
            String tempname=null;
            for(int i=0; i<files.length; i++)
            {   File temp=new File(files[i]);
                if(temp.isFile())
                if(getfilepath(files[i]).compareTo(filepath)==0)
                    name.addElement(getfilename(files[i]));
                else
                    return name;
            }
            return name;
        }

        String getpath()
        {   return filepath;
        }

        String getfilepath(String name)
        {   int index=name.lastIndexOf("\\");
            String path=name.substring(0, index+1);
            return path;
        }

        String getfilename(String name)
        {   int index=name.lastIndexOf("\\");
            String FileName=name.substring(index+1);
            return FileName;
        }
    }

    class destination
    {   String[] direct;
        destination(String[] directory)
        {   direct=directory;
        }
```

```java
    String getdirectory()
    {   String temp;
        temp=direct[direct.length-1];
        if(direct.length>2 && !(new File(temp).isDirectory()))
        {   System.out.println("The target directory doesn't exist.
                Do you want to create directory:"+temp+" (y/n)");
            int YorN=0;
            try
            {   BufferedReader in=new BufferedReader(
                    new InputStreamReader(System.in));
                YorN=in.read();
                if(YorN=='y')
                {   File tdir=new File(temp);
                    if(tdir.isFile())
                    {   System.out.println(temp+
                            " is a file but not a directory.
                            Please run this program with another target
                            directory parameter");
                        System.exit(1);
                    }
                    if(!(tdir.mkdirs()))
                    {   System.out.println("can't create "+temp);
                        System.exit(1);
                    }
                }
                else
                {   System.exit(1);
                }
            }catch(Exception e)
            {   e.printStackTrace();
            }
        }
        int length=temp.length();
        if(!temp.endsWith("\\"))
        temp=temp+"\\";
        return temp;
    }
}

class errMes
{   public void  usage()
    {   System.out.println("usage:copy [source] [destination]");
        System.exit(1);
    }
}
```

为测试程序的运行结果,进行如下三步操作:
(1) 显示当前目录下的文件列表。
(2) 执行 copy 命令。
(3) 再次显示当前目录下的文件列表。
程序的执行结果如图 11-1～图 11-3 所示。

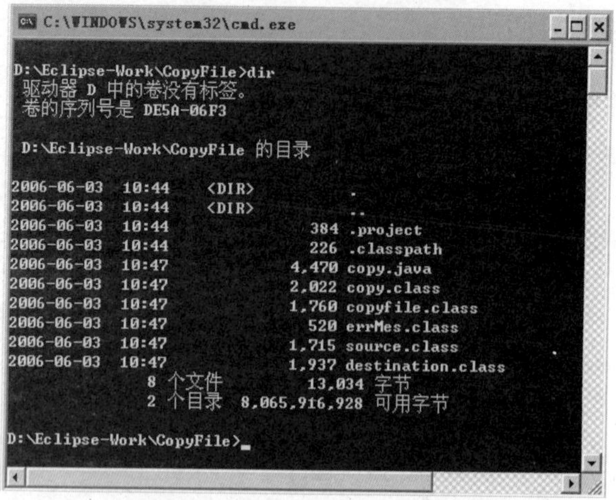

图 11-1 命令执行前的文件列表

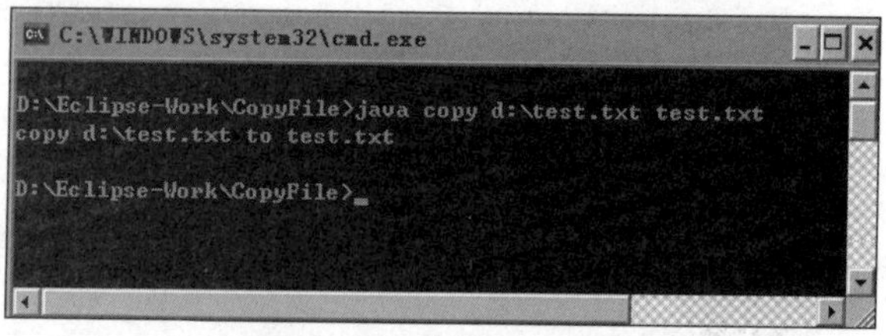

图 11-2 执行 copy 程序

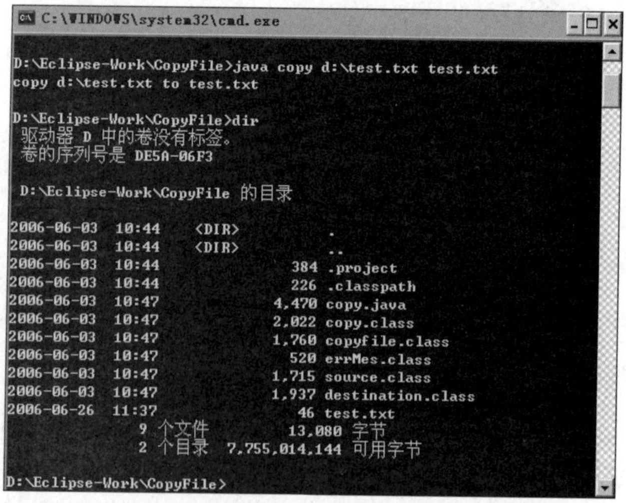

图 11-3 命令执行后的文件列表

11.7 利用输入/输出流及文件类编写一个程序，可以实现在屏幕显示文本文件的功能。程序的命令行参数的形式及操作功能均类似于 DOS 中的 type 命令，同时能够显示文件的有关属性，如文件名、路径、修改时间、文件大小等。

解：输出文件内容的命令格式为：type filename，其中，filename 是文件名，它保存在 args[0]中。

可以使用：

```
FileInputStream filestream=new FileInputStream(args[0]);
BufferedReader os=new BufferedReader(new InputStreamReader(filestream));
```

得到输入流，一行行地读入，并显示在屏幕上。

程序代码实现如下：

```java
import java.io.*;
import java.util.Date;
public class type
{   public static void main(String args[])
    {   String data=null;
        if(args.length !=1)
        {   System.out.print("usage: type filename");
            System.exit(1);
        }
        try
        {   File txtfile=new File(args[0]);
            FileInputStream filestream=new FileInputStream(args[0]);
            BufferedReader os=new BufferedReader(
                new InputStreamReader(filestream));
            for(;;)
            {   data=os.readLine();
                if(data==null)
                    break;
                System.out.println(data);
            }
            System.out.println("file's name is : "+txtfile.getName());
            System.out.println("file's path is :" +
                txtfile.getAbsolutePath());
            Date lastModifiedTime=new Date(txtfile.lastModified());
            System.out.println("file's last modifered time is : " +
                lastModifiedTime);
            System.out.println("file;s length is :"+txtfile.length() +
                "bytes");
        }catch(IOException e)
        {   e.printStackTrace();
        }
```

            }
        }

程序的执行结果如图 11-4 所示。

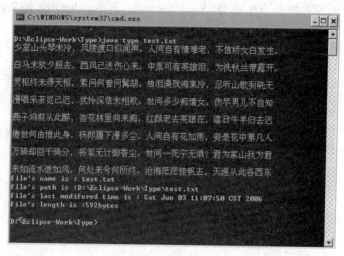

图 11-4　显示文件内容

11.8　使用缓冲区输出流的好处是什么？为什么关闭一个缓冲区输出流之前，应使用 flush() 方法？

解：缓冲区输出流在数据流上增加了一个缓冲区。当输出数据时，数据并没有直接传到与输出流相连的外设上，而是先写入缓冲区，然后将缓冲区的内容整块（块的大小可以进行设置）写入输出流中。由于采用这个办法减少了对外设的读写次数，降低了不同硬件设备之间速度的差异，提高了 I/O 操作的效率，从而对于有大量 I/O 操作的程序具有非常重要的意义。

使用缓冲区输出流时，程序的输出立即写入缓冲区中。由于当缓冲区积累一定量后才向真正的输出流中传输数据，所以在关闭缓冲区输出流时，缓冲区中可能还有未能及时写到输出流中的数据，需要使用 flush() 方法强制输出缓冲区中的剩余数据，以确保缓冲区内的所有数据全部写入输出流中。

11.9　读者和写者的作用是什么？

解：读者（Reader）和写者（Writer）是 Java 提供的对不同平台之间数据流中数据进行转换的功能。

与其他程序设计语言使用 ASCII 字符集不同，Java 使用 Unicode 字符集来表示字符串和字符。ASCII 字符集是以一个字节（8 位）来表示一个字符，所以可以认为一个字符就是一个字节（byte）。但 Java 使用的 Unicode 是一种大字符集，要用两个字节（16 位）来表示一个字符，这时字节与字符就不再统一了。为了实现与其他程序语言的交互，或是允许程序用于不同的平台，Java 必须提供一种 16 位的数据流处理方案，使得数据流中的数据可以进行与以往等效的处理。这种 16 位方案被称做读者和写者。像数据流一样，在 java.io 包中有许多的类对其进行支持。其中最重要的方案是 InputStreamReader 和

OutputStreamWriter,这两个类是字节流和读者、写者的接口。在构造一个 InputStreamReader 或 OutputStreamWriter 时,还定义了 16 位 Unicode 和其他平台的特定表示方法之间的转换规则。使用这种转换系统,Java 能够充分利用本地平台字符集设置的灵活性,同时又可通过内部使用 Unicode 而保留平台无关性。

11.10 什么叫作对象的持久化?如何实现对象的持久化?

解:把对象存为某种永久存储类型称为对象的持久化。说一个对象是可持久的,意味着可以把这个对象存入磁盘、磁带,或传入另一台机器,保存在它的内存或磁盘中。

让对象实现接口 java.io.Serializable,即可实现对象的持久化。

保存在内存中的数据在机器断电后会丢失,持久化可以使对象保存在存储设备中,以便下次需要的时候再将数据取出来。

11.11 图书馆用一个文本文件 booklist.txt 记录图书的书目,其中包括:book1、book2、…、book10。现在又采购了一批新书,请利用本章中的内容编写一个程序,将新书的书目添加到原来的文本文件中。

解:本例将设计 4 个按钮,分别是:open、save、exit 和 add,对应打开、保存、退出及添加功能。

单击 open 将打开已有书目目录。书目目录是一个文本文件,文件名为 booklist.txt,保存在你设置的目录下,本例中是 E:\java\javabook 目录。读者可以修正为本机上的实际路径名。

使用者可以在输入框内输入新书信息,单击 add 按钮后,该信息将添加到书目列表区的最后。单击 save 按钮则将新的内容保存到原文件中。如果没有单击该按钮,则文件不做任何修改,忽略刚才输入的全部信息。

单击 exit 按钮将结束程序的执行,退出系统。

程序代码实现如下:

```
import java.awt.*;
import java.awt.event.*;
import java.io.*;
import java.util.*;

public class  booklist extends Frame implements ActionListener
{    Vector savestring=new Vector();
    Frame f;
    TextField addbook;
    TextArea listbook;
    Label l1, l2;
    Button bopen, bexit, bsave, badd;              //定义 4 个按钮

    public static void main(String args[])
    {    booklist bl=new booklist();
        bl.init();
    }
```

```java
public void init()
{   f=new Frame("book list");
    f.setLayout(new FlowLayout());              //布局管理器
    //两个标签
    l1=new Label("Pls add your book here", Label.RIGHT);
    l2=new Label("Book list");
    //书名输入域
    addbook=new TextField("", 30);
    //书信息输入域
    listbook=new TextArea("", 7, 40, TextArea.SCROLLBARS_BOTH);

    bopen=new Button("open");
    bsave=new Button("save");
    bexit=new Button("exit");
    badd=new Button("add");

    //给 4 个按钮注册监听器
    bexit.addActionListener(this);
    bopen.addActionListener(this);
    bsave.addActionListener(this);
    badd.addActionListener(this);
    //将基本元素放到框架中
    f.add(l1);
    f.add(addbook);
    f.add(l2);
    f.add(listbook);
    f.add(bopen);
    f.add(bsave);
    f.add(bexit);
    f.add(badd);
    //初始时,添加按钮及保存按钮都设为不可按状态
    badd.setEnabled(false);
    bsave.setEnabled(false);
    f.setSize(300, 320);
    f.setVisible(true);
}

public void actionPerformed(ActionEvent e)
{   if(e.getSource()==bexit)                    //判定是哪个按钮
        System.exit(0);                         //退出
    if(e.getSource()==bopen)                    //打开文件按钮
    {   try{
            FileInputStream booklist=new FileInputStream(
```

```
                "E:\\java\\javabook\\booklist.txt");    //指定目录下的文件
            BufferedReader in=new BufferedReader(
                new InputStreamReader(booklist));
            for(int i=0; ;)
            {   String book=in.readLine();
                if(book==null)
                    break;
                listbook.append(book+"\n");             //在内容框内显示
                savestring.addElement(book);
            }
            badd.setEnabled(true);                      //有内容后,添加按钮可用
            bsave.setEnabled(true);                     //有内容后,保存按钮可用
            in.close();
            booklist.close();
        }catch(FileNotFoundException ee)
        {   ee.printStackTrace();
        }catch(IOException ee)
        {   ee.printStackTrace();
        }
    }
    if(e.getSource()==badd)                             //添加按钮的处理
    {   String txt=addbook.getText();
        if(txt==null)
            return;
        else
        {   listbook.append(txt+"\n");
            savestring.addElement(txt);
        }
    }
    if(e.getSource()==bsave)                            //保存按钮的处理
    {   try{
            FileOutputStream booklist=new FileOutputStream(
                "E:\\ java\\javabook\\booklist.txt");    //保存到指定目录下
            BufferedWriter out=new BufferedWriter(
                new OutputStreamWriter(booklist));
            for(int i=0; i<savestring.size(); i++)
            {   String book=savestring.elementAt(i).toString();
                out.write(book+"\n");
            }
            out.close();
            booklist.close();
        }catch(FileNotFoundException ee)
        {   ee.printStackTrace();
        }catch(IOException ee)
```

```
            {   ee.printStackTrace();
            }
        }
    }
}
```

程序执行的初始界面如图 11-5 所示。

若 booklist.txt 中保存了 10 本书,则书目信息如下:

book1 Java 算法 电子工业出版社
book2 数据结构 南开大学出版社
book3 C++ 语言 清华大学出版社
book4 网络管理教程 机械工业出版社
book5 机器人控制技术 科学出版社
book6 人工智能 清华大学出版社
book7 数据库技术 北京邮电大学出版社
book8 计算机图形学 中国人民大学出版社
book9 多媒体系统设计 高等教育出版社
book10 操作系统 清华大学出版社

单击 open 后的界面显示如图 11-6 所示。

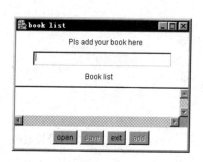

图 11-5　书目列表的初始界面

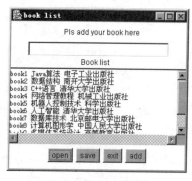

图 11-6　列出书目的效果

11.12　设计一个程序,对一个保存英文文章的文本文件进行统计,最后给出每个英文字符及每个标点的出现次数,按出现次数的降序排列。

解:定义一个类 CharSet,包含两个私有成员 keyChar 和 keyValue,分别用于存放字符及对应的出现次数。定义一个元素类型为 CharSet 的数组 ArrayList,每读入一个字符,判断 ArrayList 中是否存在该字符。若存在则将其对应的 keyValue 值加 1;若不存在则在 ArrayList 中添加一个新的 keyChar,且对应的 keyValue 值为 1。所有字符都读入 ArrayList 后,使用 Collections.sort(List list, Comparator ReverseComparator)方法按 keyValue 降序排列。其中 ReverseComparator 接口需要自行实现。

程序代码实现如下:

```
//CharSet.java
```

```java
public class CharSet                          //保存字符及对应出现次数的类
{   private char keyChar;                     //字符
    private int keyValue;                     //出现的次数

    public CharSet(char keyChar, int keyValue)
    {   this.keyChar=keyChar;
        this.keyValue=keyValue;
    }

    public char getKeyChar()                  //返回字符值
    {   return keyChar;
    }

    public void setKeyChar(char keyChar)      //设置字符值
    {   this.keyChar=keyChar;
    }

    public int getKeyValue()                  //返回出现次数
    {   return keyValue;
    }

    public void setKeyValue(int keyValue)     //设置出现次数
    {   this.keyValue=keyValue;
    }
}

//ReverseComparator.java,实现降序排列的Comparator接口
import java.io.Serializable;
import java.util.Comparator;

public final class ReverseComparator implements Comparator, Serializable
{   public int compare(Object o1, Object o2)
    {   CharSet set1=(CharSet)o1;
        CharSet set2=(CharSet)o2;
        return set2.getKeyValue()-set1.getKeyValue();
    }
}

//Statistics.java 统计字符出现频率
import java.io.BufferedReader;
import java.io.FileNotFoundException;
import java.io.FileReader;
import java.io.IOException;
import java.util.*;
```

```java
public class Statistics
{   public static final Comparator reverseComparator=new ReverseComparator();

    public static boolean isInList(List<CharSet>list, int test)
    {   for(CharSet x : list)
        {   if((int)x.getKeyChar()==test)
            {   return true;
            }
            if(Math.abs((int)x.getKeyChar()-test)==32)
            {   return true;
            }
        }
        return false;
    }

    public static void main(String[] args) throws IOException
    {   //文件名通过main方法的参数传入
        String fileName=args[0];

        try
        {   BufferedReader file=new
                BufferedReader(new FileReader(fileName));
            List<CharSet>list=new ArrayList<CharSet>();

            int next=file.read();
            while(next!=-1)
            {   if(isInList(list,next))
                {   for(CharSet x : list)
                    {   if(x.getKeyChar()==(char)next)
                        {   x.setKeyValue(x.getKeyValue()+1);
                        }
                        if(Math.abs((int)x.getKeyChar()-next)==32)
                        {   x.setKeyValue(x.getKeyValue()+1);
                        }
                    }
                }
                else
                {   if(next>=65 && next<=90)
                        list.add(new CharSet((char)(next+32),1));
                    else
                        list.add(new CharSet((char)next,1));
                }
                next=file.read();
```

```
            }
            Collections.sort(list,reverseComparator);
            for(CharSet y : list)
             {  System.out.printf("%c:%d\n",y.getKeyChar(),y.getKeyValue
());;
             }
        }
        catch(FileNotFoundException e)
        {  System.out.printf("File:%s was not found or could not be opened",
fileName);
        }
    }
}
```

11.13 设计一个通讯录,保存读者信息。通讯录中包括一般通讯录中的基本信息,也需要实现普通的检索功能。通讯录写入文件,程序执行时,需要从文件中导入数据,程序退出后再将数据保存到文件中。第一次执行时,新创建一个文件。

解:根据题意,使用文件 AddressBook.dat 保存读者信息,用类的序列化来实现读写。定义类 Address,用于保存读者信息,包含私有成员 name(姓名)、phone(固定电话)、mobile(移动电话)、address(地址)、note(备注)等。使用 ObjectInputStream、ObjectOutputStream 完成对文件的读写。界面设计使用了 awt 和 Swing,提供插入、修改、删除、查找功能。

程序代码实现如下:

```
//Address.java 类 Address 的定义
import java.io.Serializable;

public class Address implements Serializable
{   private String name;                            //姓名
    private String phone;                           //固定电话
    private String mobile;                          //移动电话
    private String address;                         //地址
    private String note;                            //备注

    public Address() { }

     public Address (String name, String phone, String mobile, String address,
     String note)
     {   this.name=name;
         this.phone=phone;
         this.mobile=mobile;
         this.address=address;
```

```java
        this.note=note;
    }

    //对类成员的相关的访问函数
    public String getAddress()
    {   return address;
    }

    public void setAddress(String address)
    {   this.address=address;
    }

    public String getMobile()
    {   return mobile;
    }

    public void setMobile(String mobile)
    {   this.mobile=mobile;
    }

    public String getName()
    {   return name;
    }

    public void setName(String name)
    {   this.name=name;
    }

    public String getNote()
    {   return note;
    }

    public void setNote(String note)
    {   this.note=note;
    }

    public String getPhone()
    {   return phone;
    }

    public void setPhone(String phone)
    {   this.phone=phone;
    }
}
```

```java
//AddressBook.java 通讯录程序实现代码
import java.awt.*;
import java.awt.event.WindowEvent;
import javax.swing.*;
import java.io.*;
import java.util.ArrayList;

public class AddressBook extends JFrame
{   private final String fileName="AddressBook.dat";
    private ArrayList<Address>addressList=new ArrayList<Address>();
    private boolean changeFlag=false;              //修改标志
    //定义窗口内必要的元素
    private JPanel jContentPane=null;
    private JMenuBar jJMenuBar=null;
    private JMenu file=null;
    private JMenu edit=null;
    private JMenu help=null;
    private JMenuItem savefile=null;
    private JMenuItem exitfile=null;
    private JMenuItem insert=null;
    private JMenuItem delete=null;
    private JMenuItem change=null;
    private JMenuItem search=null;
    private JMenuItem about=null;
    private JMenuItem helpfile=null;
    private JPanel jPanel=null;
    private JButton newitem=null;
    private JButton deleteitem=null;
    private JButton edititem=null;
    private JButton finditem=null;
    private JScrollPane jScrollPane=null;
    private JDialog insertDialog=null;
    private JPanel jContentPane1=null;
    private JTextField insertTextField=null;
    private JButton insertButton=null;
    private JLabel insertLabel=null;
    private JLabel insertLabel1=null;
    private JTextField insertTextField1=null;
    private JLabel insertLabel2=null;
    private JLabel insertLabel3=null;
    private JLabel insertLabel4=null;
    private JTextField insertTextField2=null;
    private JButton insertButton1=null;
```

```java
        private JTextField insertTextField3=null;
        private JTextField insertTextField4=null;
        private JDialog deleteDialog=null;
        private JPanel jContentPane2=null;
        private JDialog editDialog=null;
        private JPanel jContentPane3=null;
        private JDialog findDialog=null;
        private JPanel jContentPane4=null;
        private JLabel deleteLabel=null;
        private JTextField deleteTextField=null;
        private JButton deleteButton=null;
        private JButton deleteButton1=null;
        private JLabel editLabel=null;
        private JLabel editLabel1=null;
        private JLabel editLabel2=null;
        private JLabel editLabel3=null;
        private JTextField editTextField=null;
        private JTextField editTextField1=null;
        private JTextField editTextField2=null;
        private JTextField editTextField3=null;
        private JLabel editLabel4=null;
        private JTextField editTextField4=null;
        private JButton editButton=null;
        private JButton editButton1=null;
        private JLabel findLabel=null;
        private JTextField findTextField=null;
        private JButton findButton=null;
        private JButton findButton1=null;
        private JTextArea jTextArea=null;

        private JMenuBar getJJMenuBar()
        {   if (jJMenuBar==null)
            {   jJMenuBar=new JMenuBar();
                jJMenuBar.add(getFile());
                jJMenuBar.add(getEdit());
                jJMenuBar.add(getHelp());
            }
            return jJMenuBar;
        }

        private JMenu getFile()
        {   if (file==null)
            {   file=new JMenu();
                file.setText("文件");
```

```java
            file.add(getSave());
            file.add(getExitfile());
        }
        return file;
    }

    private JMenu getEdit()
    {   if (edit==null)
        {   edit=new JMenu();
            edit.setText("编辑");
            edit.add(getInsert());
            edit.add(getDelete());
            edit.add(getChange());
            edit.add(getSearch());
        }
        return edit;
    }

    private JMenu getHelp()
    {   if (help==null)
        {   help=new JMenu();
            help.setText("帮助");
            help.add(getAbout());
            help.add(getHelpfile());
        }
        return help;
    }

    private JMenuItem getSave()
    {   if (savefile==null)
        {   savefile=new JMenuItem();
            savefile.setText("保存");
            savefile.addActionListener(new java.awt.event.ActionListener()
                {   public void actionPerformed(java.awt.event.ActionEvent e)
                    {   saveFile();
                    }
                });
        }
        return savefile;
    }

    private JMenuItem getExitfile()
    {   if (exitfile==null)
        {   exitfile=new JMenuItem();
```

```java
            exitfile.setText("退出");
            exitfile.addActionListener(new java.awt.event.ActionListener()
            {   public void actionPerformed(java.awt.event.ActionEvent e)
                {   doExit();
                }
            });
        }
        return exitfile;
    }

    private JMenuItem getInsert()
    {   if (insert==null)
        {   insert=new JMenuItem();
            insert.setText("添加");
            insert.addActionListener(new java.awt.event.ActionListener()
            {   public void actionPerformed(java.awt.event.ActionEvent e)
                {   insertDialog=getInsertDialog();
                    insertDialog.setVisible(true);
                }
            });
        }
        return insert;
    }

    private JMenuItem getDelete()
    {   if (delete==null)
        {   delete=new JMenuItem();
            delete.setText("删除");
            delete.addActionListener(new java.awt.event.ActionListener()
            {   public void actionPerformed(java.awt.event.ActionEvent e)
                {   deleteDialog=getDeleteDialog();
                    deleteDialog.setVisible(true);
                }
            });
        }
        return delete;
    }

    private JPanel getJPanel()
    {   if (jPanel==null)
        {   jPanel=new JPanel();
            jPanel.add(getNewitem(), null);
            jPanel.add(getDeleteitem(), null);
            jPanel.add(getEdititem(), null);
```

```java
            jPanel.add(getFinditem(), null);
        }
        return jPanel;
    }

    private JButton getNewitem()
    {   if (newitem==null)
        {   newitem=new JButton();
            newitem.setText("添加");
            newitem.addActionListener(new java.awt.event.ActionListener()
                {   public void actionPerformed(java.awt.event.ActionEvent e)
                    {   insertDialog=getInsertDialog();
                        insertDialog.setVisible(true);
                    }
                });
        }
        return newitem;
    }

    private JButton getDeleteitem()
    {   if (deleteitem==null)
        {   deleteitem=new JButton();
            deleteitem.setText("删除");
            deleteitem.addActionListener(new java.awt.event.ActionListener()
                {   public void actionPerformed(java.awt.event.ActionEvent e)
                    {   deleteDialog=getDeleteDialog();
                        deleteDialog.setVisible(true);
                    }
                });
        }
        return deleteitem;
    }

    private JButton getEdititem()
    {   if (edititem==null)
        {   edititem=new JButton();
            edititem.setText("修改");
            edititem.addActionListener(new java.awt.event.ActionListener()
                {   public void actionPerformed(java.awt.event.ActionEvent e)
                    {   editDialog=getEditDialog();
                        editDialog.setVisible(true);
                    }
                });
        }
```

```java
        return edititem;
    }

    private JButton getFinditem()
    {   if (finditem==null)
        {   finditem=new JButton();
            finditem.setText("查找");
            finditem.addActionListener(new java.awt.event.ActionListener()
            {   public void actionPerformed(java.awt.event.ActionEvent e)
                {   findDialog=getFindDialog();
                    findDialog.setVisible(true);
                }
            });
        }
        return finditem;
    }

    private JScrollPane getJScrollPane()
    {   if (jScrollPane==null)
        {   jScrollPane=new JScrollPane();
            jScrollPane.setViewportView(getJTextArea());
        }
        return jScrollPane;
    }

    private JMenuItem getChange()
    {   if (change==null)
        {   change=new JMenuItem();
            change.setText("修改");
            change.addActionListener(new java.awt.event.ActionListener()
            {   public void actionPerformed(java.awt.event.ActionEvent e)
                {   editDialog=getEditDialog();
                    editDialog.setVisible(true);
                }
            });
        }
        return change;
    }

    private JMenuItem getSearch()
    {   if (search==null)
        {   search=new JMenuItem();
            search.setText("查找");
            search.addActionListener(new java.awt.event.ActionListener()
```

```java
            {   public void actionPerformed(java.awt.event.ActionEvent e)
                {   findDialog=getFindDialog();
                    findDialog.setVisible(true);
                }
            });
        }
        return search;
    }

    private JMenuItem getAbout()
    {   if (about==null)
        {   about=new JMenuItem();
            about.setText("版本");
        }
        return about;
    }

    private JMenuItem getHelpfile()
    {   if (helpfile==null)
        {   helpfile=new JMenuItem();
            helpfile.setText("使用方法");
        }
        return helpfile;
    }

    private JDialog getInsertDialog()
    {   if (insertDialog==null)
        {   insertDialog=new JDialog();
            insertDialog.setTitle("新建");
            insertDialog.setBounds(new java.awt.Rectangle(100,100,306,219));
            insertDialog.setContentPane(getJContentPane1());
        }
        return insertDialog;
    }

    private JPanel getJContentPane1()
    {   if (jContentPane1==null)
        {   insertLabel4=new JLabel();
            insertLabel4.setBounds(new java.awt.Rectangle(27,114,96,21));
            insertLabel4.setHorizontalAlignment(javax.swing.SwingConstants.
            RIGHT);
            insertLabel4.setText("备注:");
            insertLabel3=new JLabel();
            insertLabel3.setBounds(new java.awt.Rectangle(27,87,96,21));
```

```
            insertLabel3.setHorizontalAlignment(javax.swing.SwingConstants.
            RIGHT);
            insertLabel3.setText("地址:");
            insertLabel2=new JLabel();
            insertLabel2.setBounds(new java.awt.Rectangle(27,60,96,21));
            insertLabel2.setHorizontalAlignment(javax.swing.SwingConstants.
            RIGHT);
            insertLabel2.setText("移动电话:");
            insertLabel1=new JLabel();
            insertLabel1.setBounds(new java.awt.Rectangle(27,33,96,21));
            insertLabel1.setHorizontalAlignment(javax.swing.SwingConstants.
            RIGHT);
            insertLabel1.setText("固定电话:");
            insertLabel=new JLabel();
            insertLabel.setBounds(new java.awt.Rectangle(27,6,96,21));
            insertLabel.setHorizontalAlignment(javax.swing.SwingConstants.
            RIGHT);
            insertLabel.setText("姓名:");
            jContentPane1=new JPanel();
            jContentPane1.setLayout(null);
            jContentPane1.add(getInsertTextField(), null);
            jContentPane1.add(getInsertButton(), null);
            jContentPane1.add(insertLabel, null);
            jContentPane1.add(insertLabel1, null);
            jContentPane1.add(getInsertTextField1(), null);
            jContentPane1.add(insertLabel2, null);
            jContentPane1.add(insertLabel3, null);
            jContentPane1.add(insertLabel4, null);
            jContentPane1.add(getInsertTextField2(), null);
            jContentPane1.add(getInsertButton1(), null);
            jContentPane1.add(getInsertTextField3(), null);
            jContentPane1.add(getInsertTextField4(), null);
        }
        return jContentPane1;
    }

    private JTextField getInsertTextField()
    {   if (insertTextField==null)
        {   insertTextField=new JTextField();
            insertTextField.setBounds(new java.awt.Rectangle(150,6,134,21));
        }
        return insertTextField;
    }
```

```java
    private JButton getInsertButton()
    {   if (insertButton==null)
        {   insertButton=new JButton();
            insertButton.setText("确定");
            insertButton.setBounds(new java.awt.Rectangle(64,146,60,23));
            insertButton.addActionListener(new java.awt.event.ActionListener()
                {   public void actionPerformed(java.awt.event.ActionEvent e)
                    {   insertAddress();
                        insertDialog.setVisible(false);
                        clearInsertDialog();
                    }
                });
        }
        return insertButton;
    }

    private JTextField getInsertTextField1()
    {   if (insertTextField1==null)
        {   insertTextField1=new JTextField();
            insertTextField1.setBounds(new java.awt.Rectangle(150,33,134,21));
        }
        return insertTextField1;
    }

    private JTextField getInsertTextField2()
    {   if (insertTextField2==null)
        {   insertTextField2=new JTextField();
            insertTextField2.setBounds(new java.awt.Rectangle(150,60,134,21));
        }
        return insertTextField2;
    }

    private JButton getInsertButton1()
    {   if (insertButton1==null)
        {   insertButton1=new JButton();
            insertButton1.setBounds(new java.awt.Rectangle(188,146,60,23));
            insertButton1.setText("取消");
            insertButton1.addActionListener(new java.awt.event.ActionListener()
                {   public void actionPerformed(java.awt.event.ActionEvent e)
                    {   insertDialog.setVisible(false);
                        clearInsertDialog();
                    }
                });
        }
```

```java
        return insertButton1;
    }

    private JTextField getInsertTextField3()
    {   if (insertTextField3==null)
        {   insertTextField3=new JTextField();
            insertTextField3.setBounds(new java.awt.Rectangle(150,87,134,21));
        }
        return insertTextField3;
    }

    private JTextField getInsertTextField4()
    {   if (insertTextField4==null)
        {   insertTextField4=new JTextField();
            insertTextField4.setBounds(new java.awt.Rectangle(150,114,134,21));
        }
        return insertTextField4;
    }

    private JPanel getJContentPane()
    {   if (jContentPane==null)
        {   GridLayout gridLayout1=new GridLayout();
            gridLayout1.setRows(1);
            jContentPane=new JPanel();
            jContentPane.setLayout(new BoxLayout(getJContentPane(),
                BoxLayout.Y_AXIS));
            jContentPane.add(getJScrollPane(), null);
            jContentPane.add(getJPanel(), null);
        }
        return jContentPane;
    }

    private JDialog getDeleteDialog()
    {   if (deleteDialog==null)
        {   deleteDialog=new JDialog();
            deleteDialog.setTitle("删除");
            deleteDialog.setBounds(new java.awt.Rectangle(100,100,297,192));
            deleteDialog.setContentPane(getJContentPane2());
        }
        return deleteDialog;
    }

    private JPanel getJContentPane2()
    {   if (jContentPane2==null)
```

```java
        {   deleteLabel=new JLabel();
            deleteLabel.setBounds(new java.awt.Rectangle(31,38,77,25));
            deleteLabel.setHorizontalAlignment(javax.swing.SwingConstants.RIGHT);
            deleteLabel.setText("姓名:");
            jContentPane2=new JPanel();
            jContentPane2.setLayout(null);
            jContentPane2.add(deleteLabel, null);
            jContentPane2.add(getDeleteTextField(), null);
            jContentPane2.add(getDeleteButton(), null);
            jContentPane2.add(getDeleteButton1(), null);
        }
        return jContentPane2;
    }

    private JDialog getEditDialog()
    {   if (editDialog==null)
        {   editDialog=new JDialog();
            editDialog.setTitle("修改");
            editDialog.setBounds(new java.awt.Rectangle(100,100,307,224));
            editDialog.setContentPane(getJContentPane3());
        }
        return editDialog;
    }

    private JPanel getJContentPane3()
    {   if (jContentPane3==null)
        {   editLabel4=new JLabel();
            editLabel4.setBounds(new java.awt.Rectangle(23,124,96,21));
            editLabel4.setHorizontalAlignment(javax.swing.SwingConstants.RIGHT);
            editLabel4.setText("备注:");
            editLabel3=new JLabel();
            editLabel3.setBounds(new java.awt.Rectangle(23,95,96,21));
            editLabel3.setHorizontalAlignment(javax.swing.SwingConstants.RIGHT);
            editLabel3.setText("地址:");
            editLabel2=new JLabel();
            editLabel2.setBounds(new java.awt.Rectangle(23,66,96,21));
            editLabel2.setHorizontalAlignment(javax.swing.SwingConstants.RIGHT);
            editLabel2.setText("移动电话:");
            editLabel1=new JLabel();
            editLabel1.setBounds(new java.awt.Rectangle(23,37,96,21));
            editLabel1.setHorizontalAlignment(javax.swing.SwingConstants.RIGHT);
            editLabel1.setText("固定电话:");
            editLabel=new JLabel();
            editLabel.setBounds(new java.awt.Rectangle(23,8,96,21));
```

```java
            editLabel.setHorizontalAlignment(javax.swing.SwingConstants.RIGHT);
            editLabel.setText("姓名:");
            jContentPane3=new JPanel();
            jContentPane3.setLayout(null);
            jContentPane3.add(editLabel, null);
            jContentPane3.add(editLabel1, null);
            jContentPane3.add(editLabel2, null);
            jContentPane3.add(editLabel3, null);
            jContentPane3.add(getEditTextField(), null);
            jContentPane3.add(getEditTextField1(), null);
            jContentPane3.add(getEditTextField2(), null);
            jContentPane3.add(getEditTextField3(), null);
            jContentPane3.add(editLabel4, null);
            jContentPane3.add(getEditTextField4(), null);
            jContentPane3.add(getEditButton(), null);
            jContentPane3.add(getEditButton1(), null);
        }
        return jContentPane3;
    }


    private JDialog getFindDialog()
    {   if (findDialog==null)
        {   findDialog=new JDialog();
            findDialog.setTitle("查询");
            findDialog.setBounds(new java.awt.Rectangle(100,100,295,206));
            findDialog.setContentPane(getJContentPane4());
        }
        return findDialog;
    }


    private JPanel getJContentPane4()
    {   if (jContentPane4==null)
        {   findLabel=new JLabel();
            findLabel.setBounds(new java.awt.Rectangle(29,43,81,25));
            findLabel.setHorizontalAlignment(javax.swing.SwingConstants.RIGHT);
            findLabel.setText("姓名:");
            jContentPane4=new JPanel();
            jContentPane4.setLayout(null);
            jContentPane4.add(findLabel, null);
            jContentPane4.add(getFindTextField(), null);
            jContentPane4.add(getFindButton(), null);
            jContentPane4.add(getFindButton1(), null);
        }
        return jContentPane4;
```

```java
        }

    private JTextField getDeleteTextField()
    {   if (deleteTextField==null)
        {   deleteTextField=new JTextField();
            deleteTextField.setBounds(new java.awt.Rectangle(114,38,150,25));
        }
        return deleteTextField;
    }

    private JButton getDeleteButton()
    {   if (deleteButton==null)
        {   deleteButton=new JButton();
            deleteButton.setBounds(new java.awt.Rectangle(45,101,77,24));
            deleteButton.setText("确定");
            deleteButton.addActionListener(new java.awt.event.ActionListener()
            {   public void actionPerformed(java.awt.event.ActionEvent e)
                {   deleteAddress(deleteTextField.getText());
                    deleteDialog.setVisible(false);
                    clearDeleteDialog();
                }
            });
        }
        return deleteButton;
    }

    private JButton getDeleteButton1()
    {   if (deleteButton1==null)
        {   deleteButton1=new JButton();
            deleteButton1.setBounds(new java.awt.Rectangle(167,101,77,24));
            deleteButton1.setText("取消");
            deleteButton1.addActionListener(new java.awt.event.ActionListener()
            {   public void actionPerformed(java.awt.event.ActionEvent e)
                {   deleteDialog.setVisible(false);
                    clearDeleteDialog();
                }
            });
        }
        return deleteButton1;
    }

    private JTextField getEditTextField()
    {   if (editTextField==null)
        {   editTextField=new JTextField();
```

```java
            editTextField.setBounds(new java.awt.Rectangle(142,8,134,21));
        }
        return editTextField;
    }

    private JTextField getEditTextField1()
    {   if (editTextField1==null)
        {   editTextField1=new JTextField();
            editTextField1.setBounds(new java.awt.Rectangle(142,37,134,21));
        }
        return editTextField1;
    }

    private JTextField getEditTextField2()
    {   if (editTextField2==null)
        {   editTextField2=new JTextField();
            editTextField2.setBounds(new java.awt.Rectangle(142,66,134,21));
        }
        return editTextField2;
    }

    private JTextField getEditTextField3()
    {   if (editTextField3==null)
        {   editTextField3=new JTextField();
            editTextField3.setBounds(new java.awt.Rectangle(142,95,134,21));
        }
        return editTextField3;
    }

    private JTextField getEditTextField4()
    {   if (editTextField4==null)
        {   editTextField4=new JTextField();
            editTextField4.setBounds(new java.awt.Rectangle(142,124,134,21));
        }
        return editTextField4;
    }

    private JButton getEditButton()
    {   if (editButton==null)
        {   editButton=new JButton();
            editButton.setBounds(new java.awt.Rectangle(59,153,60,23));
            editButton.setText("确定");
            editButton.addActionListener(new java.awt.event.ActionListener()
            {   public void actionPerformed(java.awt.event.ActionEvent e)
```

```
                    {   editAddress(editTextField.getText());
                        editDialog.setVisible(false);
                        clearEditDialog();
                    }
                });
        }
        return editButton;
    }

    private JButton getEditButton1()
    {   if (editButton1==null)
        {   editButton1=new JButton();
            editButton1.setBounds(new java.awt.Rectangle(178,153,60,23));
            editButton1.setText("取消");
            editButton1.addActionListener(new java.awt.event.ActionListener()
                {   public void actionPerformed(java.awt.event.ActionEvent e)
                    {   editDialog.setVisible(false);
                        clearEditDialog();
                    }
                });
        }
        return editButton1;
    }

    private JTextField getFindTextField()
    {   if (findTextField==null)
        {   findTextField=new JTextField();
            findTextField.setBounds(new java.awt.Rectangle(114,43,150,25));
        }
        return findTextField;
    }

    private JButton getFindButton()
    {   if (findButton==null)
        {   findButton=new JButton();
            findButton.setBounds(new java.awt.Rectangle(59,111,60,23));
            findButton.setText("确定");
            findButton.addActionListener(new java.awt.event.ActionListener()
                {   public void actionPerformed(java.awt.event.ActionEvent e)
                    {   findAddress(findTextField.getText());
                        findDialog.setVisible(false);
                        clearFindDialog();
                    }
                });
```

```java
        }
        return findButton;
    }

    private JButton getFindButton1()
    {   if (findButton1==null)
        {   findButton1=new JButton();
            findButton1.setBounds(new java.awt.Rectangle(163,111,60,23));
            findButton1.setText("取消");
            findButton1.addActionListener(new java.awt.event.ActionListener()
            {   public void actionPerformed(java.awt.event.ActionEvent e)
                {   findDialog.setVisible(false);
                    clearFindDialog();
                }
            });
        }
        return findButton1;
    }

    private JTextArea getJTextArea()
    {   if (jTextArea==null)
        {   jTextArea=new JTextArea();
        }
        return jTextArea;
    }

    private void initialize()
    {   this.setJMenuBar(getJJMenuBar());
        this.setContentPane(getJContentPane());
        this.setTitle("通讯录");
        this.setBounds(new java.awt.Rectangle(100,100,528,259));
    }

    /***显示地址簿***/
    private void showAddressBook()
    {   jTextArea.setText("姓名\t"+"固定电话\t"+"移动电话\t"+"地址\t"+"备注\n");

        for(Address a : addressList){
            if(!(a.getName().equals("deleted")))
            {   String showAddress=a.getName() +"\t" +a.getPhone() +"\t"+
                    a.getMobile() +"\t" +a.getAddress() +"\t"+a.getNote()+"\n";
                jTextArea.append(showAddress);
            }
        }
```

```
     }

   /***添加新的地址项***/
   private void insertAddress()
   {   Address address=new Address(insertTextField.getText(),
          insertTextField1.getText(),insertTextField2.getText(),
          insertTextField3.getText(),insertTextField4.getText());
       addressList.add(address);
       changeFlag=true;
       showAddressBook();
   }

   private void clearInsertDialog()
   {   insertTextField.setText(null);
       insertTextField1.setText(null);
       insertTextField2.setText(null);
       insertTextField3.setText(null);
       insertTextField4.setText(null);
   }

   /***根据姓名删除地址项***/
   private void deleteAddress(String name)
   {   for(Address a : addressList){
          if(a.getName().equals(name))
          {   a.setName("deleted");
          }
       }
       changeFlag=true;
       showAddressBook();
   }

   private void clearDeleteDialog()
   {   deleteTextField.setText(null);
   }

   /***编辑地址项***/
   private void editAddress(String name)
   {   for(Address a : addressList)
       {   if(a.getName().equals(name))
          {   a.setAddress(editTextField3.getText());
              a.setMobile(editTextField2.getText());
              a.setPhone(editTextField1.getText());
              a.setNote(editTextField4.getText());
          }
```

```
            }
            changeFlag=true;
            showAddressBook();
    }

    private void clearEditDialog()
    {   editTextField.setText(null);
        editTextField1.setText(null);
        editTextField2.setText(null);
        editTextField3.setText(null);
        editTextField4.setText(null);
    }

    /***根据姓名查找地址项***/
    private void findAddress(String name)
    {   jTextArea.setText("姓名\t"+"固定电话\t"+"移动电话\t"+"地址\t"+"备注\n");
        for(Address a : addressList)
        {   if(a.getName().equals(name))
            {   String showAddress=a.getName()+"\t"+a.getPhone()+"\t"+a.getMobile()
                    +"\t"+a.getAddress()+"\t"+a.getNote()+"\n";
                jTextArea.append(showAddress);
            }
        }
    }

    private void clearFindDialog()
    {   findTextField.setText(null);
    }

    /***读取文件AddressBook.dat***/
    private void readFile()
    {   try
        {   ObjectInputStream ois=new ObjectInputStream(new BufferedInputStream(
                new FileInputStream(fileName)));
            Address address;
            do{
                address=(Address)ois.readObject();
                addressList.add(address);
            }while(address !=null);
        }catch(Exception e)
        {   e.getStackTrace();
        }
    }
```

```java
/***保存文件***/
private void saveFile()
{   try
    {   ObjectOutputStream oos=new ObjectOutputStream(new BufferedOutputStream(
            new FileOutputStream(fileName)));
        for(Address a : addressList)
        {   if(!(a.getName().equals("deleted")))
            {   oos.writeObject(a);
            }
        }
        oos.close();
    }catch(Exception e)
    {   e.getStackTrace();
    }
}

/***退出方法***/
private void doExit()
{   if(changeFlag==true)
    {   saveFile();
    }
    System.exit(0);
}

protected void processWindowEvent(WindowEvent e)
{   super.processWindowEvent(e);
    if(e.getID()==WindowEvent.WINDOW_CLOSING)
    {   doExit();
    }
}

public AddressBook()
{   super();
    initialize();
    readFile();
    showAddressBook();
}

public static void main(String[] args)
{   AddressBook addressBook=new AddressBook();
    addressBook.setVisible(true);
```

```
        addressBook.showAddressBook();
    }
}
```

运行初始界面如图 11-7 所示。

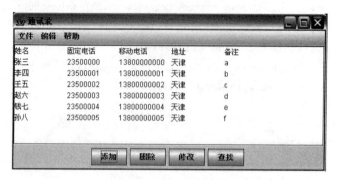

图 11-7　运行初始界面

添加地址信息界面如图 11-8 所示。

图 11-8　添加地址信息界面

添加信息后的情况如图 11-9 所示。

图 11-9　添加信息后的情况

第12章 线　　程

12.1　什么叫线程？什么叫多线程？Java的多线程有何特点？

解：线程是进程执行过程中产生的多条执行线索，是比进程更小的执行单位，它在形式上同进程十分相似——都是用一个顺序执行的语句序列来完成特定的功能。不同的是，它没有入口，也没有出口，因此其自身不能自动运行，而必须栖身于某一进程之中，由进程触发执行。在系统资源的使用上，属于同一进程的所有线程共享该进程的系统资源，但是线程之间切换的速度比进程切换要快得多。

Java中的线程包含三个主要部分，第一是虚拟CPU本身，第二是CPU执行的代码，第三是代码操作的数据。在Java中，虚拟CPU体现于Thread类中。当一个线程被构造时，它由构造方法参数、执行代码、操作数据来初始化。这三方面是各自独立的。一个线程所执行的代码与其他线程可以相同也可以不同，一个线程访问的数据与其他线程可以相同也可以不同。

目前大多数操作系统支持多任务，每一个任务就是一个进程。一个进程在其执行过程中，可以产生多个线程，形成多条执行线索，即同时存在几个执行体，按几条不同的执行路线共同工作。Java虚拟机允许一个应用程序同时有多个线程在运行。

Java语言的多线程的最大特点是它内置对多线程的支持，Java语言把线程或执行环境（execution context）当作一个封装对象，包含CPU及自己的程序代码和数据，由虚拟机提供控制。Java类库中的类java.lang.Thread允许创建并控制所创建的线程。

每个Java程序都有一个默认的主线程，分为两种情况。对于应用程序来说，主线程是main()方法执行的线索；对于Applet来说，主线程令浏览器加载并执行小应用程序。

12.2　什么叫作线程的生命周期？线程的一个生命周期包括哪些状态？各状态之间是如何进行转换的？

解：从线程的创建直到线程退出运行称为线程的生命周期。线程的一个生命周期包括以下一些状态：新建、就绪、运行、阻塞和死亡。线程状态及状态转换如图12-1所示。

一个线程被创建时，处于新建状态。一般地，新建的线程并没有立刻运行，必须通过start()方法来启动。启动后，线程进入就绪队列中，等待CPU，此时线程处于就绪状态。获得CPU的线程将被调度执行，执行自己的run()方法，此时的线程处于运行状态。正在执行的线程由于某种原因暂停执行，它将让出CPU，允许其他进程获得CPU。暂停执行的原因可能是run()方法中调用了暂停执行的方法如sleep()或yield()，也可能是线程需要执行I/O操作，或是具有更高优先级的线程抢占了CPU等。此时线程转为阻塞状态，进入阻塞队列中排队。当引起阻塞的原因消除后，线程转入就绪队列中成为就绪状态。当线程完成了全部任务或是被强制终止时，线程进入死亡状态。死亡的线程不能再被执行。

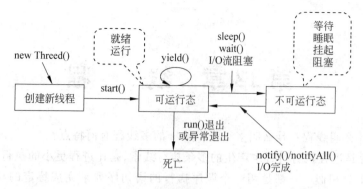

图 12-1 线程状态及状态转换

12.3 有几种创建线程的方法？分别是什么？

解：创建线程有两种方法，第一种是继承 Thread 类，定义它的一个子类，这个子类就可以表示线程。

用 Thread 类的子类创建线程的过程包括以下三步：

(1) 从 Thread 类派生出一个子类，在类中一定要实现 run()方法。

(2) 用该类创建一个对象。

(3) 用 start()方法启动线程。

创建线程的另一种方法是实现 Runnable 接口。Runnable 接口中只定义了一个方法就是 run()方法，也就是线程体。用 Runnable()接口实现多线程时，必须实现 run()方法，也需用 start()启动线程。

12.4 Thread 类中包含了哪些基本的方法？为什么 run()方法被称为线程体？

解：Java JDK 8 中提供的 Thread 类的方法列在表 11-1 中。

表 11-1 Thread 类中包含的方法

返回值	方法	功能
static int	activeCount()	返回当前线程的线程组内的活动线程数
void	checkAccess()	判定当前运行的线程是否能修改本线程
static Thread	currentThread()	返回指向当前运行线程对象的引用
static void	dumpStack()	打印当前线程的栈的变化情况
static int	enumerate(Thread[] tarray)	将当前线程的线程组及子组内的每个活动线程复制到指定数组内
static Map<Thread, StackTraceElement[]>	getAllStackTraces()	返回所有活动线程的栈的变化情况
ClassLoader	getContextClassLoader()	返回本线程的上下文类装载器
static Thread.Uncaught-ExceptionHandler	getDefaultUncaughtException-Handler()	当线程因未捕获的异常导致中断时，返回默认的处理程序
long	getId()	返回线程的标识符
String	getName()	返回线程名

续表

返回值	方法	功能
int	getPriority()	返回线程的优先级
StackTraceElement[]	getStackTrace()	返回线程栈的变化情况对应的数组
Thread.State	getState()	返回线程的状态
ThreadGroup	getThreadGroup()	返回本线程所属的线程组
Thread.UncaughtExceptionHandler	getUncaughtExceptionHandler()	当线程因未捕获的异常导致中断时,返回处理程序
static boolean	holdsLock(Object obj)	当且仅当当前线程持有指定对象的监视锁时,返回真
void	interrupt()	中断线程
static boolean	interrupted()	测试线程是否被中断
boolean	isAlive()	测试线程是否为活动的
boolean	isDaemon()	测试线程是否为守护线程
boolean	isInterrupted()	测试线程是否被中断
void	join()	等待线程死亡
void	join(long millis)	等待一段时间后线程死亡,时间最多为 millis(毫秒)
void	join(long millis, int nanos)	等待一段时间后线程死亡,时间最多为 millis(毫秒)加 nanos(十亿分之一秒)
void	run()	如果线程是用一个独立的 Runnable 接口对象创建的,则调用 Runnable 接口对象的 run()方法,否则这个方法为空并返回
void	setContextClassLoader(ClassLoader cl)	为线程设置上下文类装载器
void	setDaemon(boolean on)	标记线程是守护线程或是用户线程
static void	setDefaultUncaughtExceptionHandler(Thread.UncaughtExceptionHandler eh)	为线程设置因未捕获的异常导致中断且未定义其他处理程序时默认的处理程序
void	setName(String name)	以字符串参数 name 为线程设置名字
void	setPriority(int newPriority)	改变线程的优先级
void	setUncaughtExceptionHandler(Thread.UncaughtExceptionHandler eh)	为线程设置因未捕获的异常导致中断时默认的处理程序
static void	sleep(long millis)	让当前运行的线程休眠,时间由参数 millis(毫秒)确定
static void	sleep(long millis, int nanos)	让当前运行的线程休眠,时间由参数 millis(毫秒)及 nanos(十亿分之一秒)确定
void	start()	本线程运行,JVM 调用线程的 run()方法
String	toString()	返回本线程的字符串表示,包括线程名、优先级及线程组
static void	yield()	暂停当前正在运行的线程,允许其他线程运行

其中，start()（线程的启动）、sleep()（线程的休眠）、join()（等待线程死亡）、yield()（暂停当前线程的执行）、run()（线程体即线程的运行）等是一些基本的方法。

定义一个线程，实际上就是实现一个重要的方法即 run()方法。有两种方法可以创建线程，分别是继承 Thread 类及实现 Runnable 接口。一个线程被创建后，不论是用两种方法中的哪一种创建的，都要先调用线程对象的 run()方法。该方法是整个线程的核心，线程所要完成的任务的相关代码都定义在 run()方法中，因此一般将它称为线程体。不同的线程之间的功能差异表现为线程体的不同。

12.5 选择创建线程的方法时，应注意考虑哪些因素？

解：创建线程的两种方法其差别在于，一个是继承 Thread 类，另一个是实现 Runnable 接口。因此选择创建线程的方法时，也要着重考虑这两方面的因素。

因为 Java 只允许单重继承，如果一个类已经继承了 Thread，就不能再继承其他类；因此在需要继承其他类的情况下，就必须采用实现 Runnable 接口的创建方法。比如对于 Applet 程序，由于必须继承 java.applet.Applet，因此只能采取这种实现接口的方法。另外，有些程序已经采用了实现 Runnable 接口的方法，有时出于保持程序风格的一致性也将继续使用这种方法。

另一方面，使用 Thread 类的子类时，程序的代码可以简洁一些。此时 run()方法中的 this 实际上引用的是控制当前运行系统的 Thread 实例。对于比较简单的程序，可以考虑采用这种继承模式；但如果程序比较复杂，特别是继承关系复杂时，这种简单的继承模式将会在以后的继承中造成麻烦。

12.6 Java 是如何对线程进行调度的？

解：在 Java 中，线程调度通常是抢占式，而不是时间片式。抢占式调度是指可能有多个线程准备运行，但只有一个在真正运行。若一个线程获得了执行权，这个线程就将持续运行下去，直到它运行结束或因为某种原因而阻塞，或者有另一个高优先级线程就绪。最后一种情况称为低优先级线程被高优先级线程所抢占。

系统内部的所有线程组成两个队列，一个是就绪队列，一个是阻塞队列。

一个线程被阻塞的原因是多种多样的，可能是因为执行了 Thread.sleep()调用，故意让它暂停一段时间；也可能是因为需要等待一个较慢的外部设备，例如磁盘；还可能是等待用户的输入。所有被阻塞的线程按次序排列，放入阻塞队列中。

所有就绪（满足所有运行条件，只等待 CPU 空闲）但没有运行的线程则根据其优先级排入就绪队列中。当 CPU 空闲时，如果就绪队列不空，就绪队列中具有最高优先级的第一个线程将运行。当一个线程被抢占而停止运行时，它的运行态被改变为就绪态并放到就绪队列的队尾；同样，一个被阻塞（可能因为睡眠或等待 I/O 设备）的线程就绪后，也就是睡眠时间到或是需要的 I/O 设备已满足，通常也放到就绪队列的队尾。

12.7 方法 sleep()和 yield()都能够暂时停止当前线程的运行，并让出 CPU。两者的区别是什么？

解：sleep()是使线程暂时停止执行，放弃 CPU，进入休眠期。休眠期的长短可以由程序员指定，休眠期过后重新被调度。

而 yield()是使当前运行的线程放弃 CPU，并允许其他线程运行。yield()虽然让线程放弃了 CPU,可是由于它的优先级比较高,所以它还是可能在放弃 CPU 后马上再得到 CPU 继续运行。

12.8 为什么在多线程系统中要引入同步机制? Java 是如何实现同步机制的?

解:多线程访问共享数据时,可能会出现访问了不该访问的数据或是漏过需要访问的数据这种情况,这是因为缺乏线程和共享数据之间的协调管理机制而导致的。

举一个例子。两个线程 SENDER 和 CATCHER 同时对一个单元 ADDRESS 进行操作,每次 SENDER 向 ADDRESS 中送入一个数据,CATCHER 从 ADDRESS 中取出一个数据。只有当 SENDER 送入数据后,CATCHER 才能从中取出。如果 SENDER 没有及时送入数据,CATCHER 将取不到数据或是还取回旧数据。相反,如果 CATCHER 不及时取出 ADDRESS 中的数据,SENDER 送入的数据将覆盖 ADDRESS 中原来的数据,从而发生丢失数据的情况。

为了解决类似的问题,Java 使用对象锁的概念,为可能同时访问共享数据的代码实例定义了一个"锁定标志",使用 synchronized 关键字进行同步,它提供了操作锁定标志的方法。

使用 synchronized 为访问共享数据的代码加标志,它括住的代码即是需要同步的语句。当线程执行到被同步的语句时,将传递的对象参数设为锁定标志,禁止其他线程对该对象的访问。这时如果另一个线程企图执行 synchronized(this)中的语句,它将从对象 this 中索取锁定标记。因为这个标记不可得,故第二个线程不能继续执行。实际上这个线程将加入一个等待队列,这个等待队列与对象锁定标志相连。当标志被返还给对象时,第一个等待它的线程将得到它并继续运行。

12.9 什么叫作死锁? Java 的同步机制是否可以防止死锁的发生?

解:在多线程竞争使用资源的程序中,可能会发生这样的情况:一个线程等待另一个线程所持有的锁,而第二个线程又在等待第一个线程所持有的锁,也就是它们互相等待对方所持有的锁。一般来讲,每个线程都占有部分资源,而等待其他线程占有的另外部分资源。因为每个线程都得不到运行所需的全部资源,所以每个线程都不能继续运行。这就出现了死锁。死锁是指多个线程相互等待,使得程序无法继续运行。

Java 既不监测也没有采取办法防止死锁的出现。它的同步机制不能有效地防止死锁,死锁现象需要程序员仔细编程才能防止。例如,将资源进行编号,并控制程序按序得到资源,只有已经得到小编号的资源后,才有资格获取大编号的资源。

12.10 领会 Java 中的交互机制,并利用前面学习的内容,完成下面的程序:创建两个 Thread 子类,第一个 run()方法用于最开始的启动,并捕获第二个 Thread 对象的句柄,然后调用 wait()。第二个类的 run()方法应在过几秒以后为第一个线程调用 notifyAll(),使第一个线程能打印出一条消息。

解:根据题意,创建两个线程,一个是 ft,另一个是 st;并调用 ft.start()及 st.start(),启动两个线程。第一个线程启动后进入无限期的等待状态。第二个线程启动后,先睡眠一段时间,然后唤醒第一个线程。第一个线程被唤醒后,输出一条信息。

程序代码实现如下:

```
import java.lang.*;
```

```java
import java.io.*;

public class ThreadTest
{   public static void main(String args[])
    {   Object dummy=new Object();
        FirstThread ft=new FirstThread( dummy);
        SecondThread st=new SecondThread(dummy);
        ft.start();
        st.start();
    }
}

class FirstThread extends Thread
{   Object dummy;

    public FirstThread(Object s)
    {   dummy=s;
        System.out.println("The first thread was constructed");
    }

    public void run()
    {   try
        {   System.out.println("The first thread has been started");
            System.out.println("The first thread is waiting for
                the second one");
            synchronized(dummy)
            {   dummy.wait();
            }
            System.out.println("The second thread has made me wake up ");
        }catch(Exception e){}
    }
}

class SecondThread extends Thread
{   Object dummy;

    public SecondThread(Object s)
    {   dummy=s;
        System.out.println("The second thread was constructed");
    }

    public void run()
    {   try
```

```
        {   System.out.println("The second thread has been started");
            Thread.sleep(5000);
            synchronized(dummy)
            {   dummy.notifyAll();
            }
        }catch(Exception e){}
    }
}
```

程序的执行结果如图 12-2 所示。

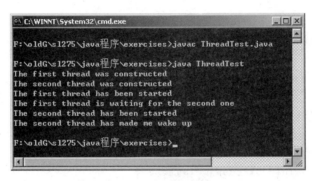

图 12-2 多线程示例

12.11 利用多线程技术和图像显示技术来完成动画设计。

提示：动画的本质就是运动的图形，即在屏幕上高速显示一系列帧连续不断的、相类似，但又有所不同的图像，从而造成动画的感觉。帧更新速率越快，图形闪动越小，动画的效果也就越好。

Java 中支持的图形格式只有 GIF 和 JPEG 两种文件格式。处理动画时涉及的主要有：java.applet 包中的 Applet 类里的两个方法：getImage(URL url)和 getImage(URL url, String name)方法，java.awt 包中的 Graphics 类的几个 drawImage()方法。

解：定义一个 Image 数组，使用 getImage(URL, String name)方法装载相应图像文件到该数组中。在 paint()方法中，使用 drawImage()方法画出图像，然后定义一个线程，让该线程睡眠一段时间，再切换成下一幅图像。线程创建通过实现 Runnable 接口的方法。

程序代码实现如下：

```
//Animation.java
import java.applet.Applet;
import java.awt.*;

public class Animation extends Applet implements Runnable
{   private Thread animation=null;
    private final int delay=500;
    private final int imageNum=5;
```

```java
        private Image[] images=new Image[imageNum];

        public void run() { }

        public void start()
        {   animation=new Thread(this);
            animation.start();
        }

        public void stop()
        {   if(animation !=null)
                animation=null;
        }

        public void paint(Graphics g)
        {   for(int i=0;i<imageNum;i++)
            {   g.drawImage(images[i],25,25,100,100,this);
                try
                {   Thread.sleep(delay);
                }catch(InterruptedException e){ }
            }
        }

        public void update(Graphics g)
        {   paint(g);
        }

        public void init()
        {   this.setSize(300, 200);
            for(int i=0;i<imageNum;i++)
            {   images[i]=getImage(getCodeBase(),(i+1)+".jpg");
            }
        }
    }
```

对应的 HTML 文档如下：

```
<html>
<applet code=Animation.class width=300 height=300>
</applet>
</html>
```

分别存放 5 个图片，放置数字 1~5。程序的执行结果如图 12-3~图 12-7 所示。

图 12-3 动画显示效果(1)

图 12-4 动画显示效果(2)

图 12-5 动画显示效果(3)

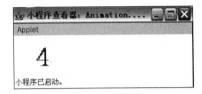

图 12-6 动画显示效果(4)

图 12-7 动画显示效果(5)

第13章　Java 的网络功能

13.1 Java 所能够提供的网络功能按层次及使用方法可以分为几类？简要说明每种方法的特点。

解：Java 所能够提供的网络功能按层次及使用方法可分为三大类。

(1) URL

这种方法是通过 URL 的网络资源表达形式确定数据在网络中的位置，利用 URL 对象中提供的相关方法，直接读入网络中的数据，或者将本地数据传送到网络的另一端。

(2) socket

socket 是指两个程序在网络上的通信连接。由于在 TCP/IP 协议下的客户/服务器软件通常使用 socket 来进行信息交流，因此这种方法也是传统网络程序经常用到的一种方式。

(3) Datagram

前两种方法都是面向连接方式的通信，其前提是程序所使用的通信管道是安全而稳定的。但是在网络条件较为复杂的情况下，这种要求未必能够达到。这时可以使用 Datagram 方式。Datagram 方式是三种网络功能中最低级的一种，它是一种面向非连接的、以数据报方式工作的通信，适用于网络状况不可靠环境下的数据传输和访问。

13.2 什么叫 socket？它在网络通信中的作用是什么？

解：socket 是为了进行网络通信引入的抽象概念，一般称作套接字或套接口。它是 TCP/IP 协议的编程接口，提供了一组 API 函数，使用这些函数就可以实现 TCP/IP 协议。socket 可以完成网上建立连接、数据传输的任务。它可以接收请求，也可以发出请求。

13.3 怎样建立 socket 连接？

解：使用 socket 方式进行建立连接的过程为：

(1) 先在服务器端生成一个 ServerSocket 实例对象，随时监听客户端的连接请求。

(2) 当客户端需要连接时，相应地要生成一个 Socket 实例对象，并发出连接请求。其中 host 参数指明该主机名，port# 参数指明该主机端口号，也即要打开 socket 接口。

(3) 服务器端通过 accept() 方法接收到客户端的请求后，开辟一个接口与之进行连接，并生成所需的 I/O 数据流。这就是将特定的输入/输出流连接到打开的 socket。

(4) 客户端和服务器端的通信都是通过一对 InputStream 和 OutputStream 进行的，然后按实际需要对 socket 进行读写操作。

(5) 通信结束后，两端分别关闭对应的 socket 接口。

13.4 在通信结束时为什么要关闭 socket？如何关闭 socket？

解：基于 socket 的通信是基于连接的通信。它可以确保整个通信过程准确无误，但是保持连接需要占用系统的内存资源。当通信结束时，socket 将不再有用，关闭它是为了节约宝贵的系统资源。

关闭 socket 之前要先关闭与之相连的输入/输出流，然后再关闭 socket。例如，在如下的程序中，数据流是 s1In 和 dis，建立 socket 为 s1。当数据传输完毕时，先关闭数据流：

```
dis.close();
s1In.close();
```

再关闭 socket：

```
s1.close();
```

示例如下：

```java
public class SimpleClient{
    public static void main(String args[]) throws IOException{
        int c;
        Socket s1;
        InputStream s1In;
        DataInputStream dis;

        //在端口 5432 打开连接
        s1=new Socket("subbert",5432);
        //获得 socket 端口的输入句柄，并从中读取数据
        s1In=s1.getInputStream();
        dis=new DataInputStream(s1In);

        String st=new String(dis.readUTF());
        System.out.println(st);

        //操作结束，关闭数据流及 socket 连接
        dis.close();
        s1In.close();
        s1.close();
    }
}
```

13.5 如何创建 URL 对象？

解：使用 URL 类的构造方法可以创建 URL 对象，JDK 8 中 URL 的构造方法主要有以下 6 种，如表 13-1 所示。

表 13-1 构造方法及解释

构造方法	解释
URL(String spec)	创建一个以字符串 spec 表示的 URL 对象
URL(String protocol, String host, int port, String file)	创建一个 URL 对象,构造方法中需要给出使用的传输协议、主机名、端口号及文件名
URL(String protocol, String host, int port, String file, URLStreamHandler handler)	创建一个 URL 对象,构造方法中需要给出使用的传输协议、主机名、端口号、文件名及句柄
URL(String protocol, String host, String file)	创建一个 URL 对象,构造方法中需要给出使用的传输协议、主机名和文件名
URL(URL context, String spec)	由给定的上下文及 spec 创建一个 URL 对象
URL(URL context, String spec, URLStreamHandler handler)	由给定的上下文、spec 及句柄创建一个 URL 对象

13.6 如何利用 URL 来读取网络资源?

解:可以通过类 URL 所提供的方法来获取对象属性,如表 13-2 所示。

表 13-2 方法及解释

方法	解释	方法	解释
String getProtocol()	获取传输协议	String getFile()	获取资源文件名称
String getHost()	获取机器名称	String getRef()	获取参考点
String getPort()	获取通信端口号		

URL 类中定义了 openStream() 方法用以读取 URL 位置的数据,其返回值是一个 InputStream 数据流。通过这个数据流对象可以进行数据的操作。

13.7 利用前面所学习的多线程的内容,改进本章中的程序,使之能够服务于多个客户。

解:根据题意,建立一个端口为 8189 的服务器,它无限期等待直到有客户连接这个端口。若客户在网上发送正确的请求连接了这个端口,则服务器与客户之间就建立一个可靠的连接。

服务器端的程序如下:

```
import java.io.*;
import java.net.*;

public class ThreadedEchoServer
{   public static void main(String[] args )
    {   int i=1;
        try
        {   ServerSocket s=new ServerSocket(8189);
            for ( ; ; )
            {   Socket incoming=s.accept();
                System.out.println("Spawning "+i);
```

```java
                new ThreadedEchoHandler(incoming, i).start();
                i++;
            }
        }catch (Exception e)
        {   System.out.println(e);
        }
    }
}

class ThreadedEchoHandler extends Thread
{   public ThreadedEchoHandler(Socket i, int c)
    {   incoming=i; counter=c;
}

    public void run()
    {   try
        {   BufferedReader in=new BufferedReader
                (new InputStreamReader(incoming.getInputStream()));
            PrintWriter out=new PrintWriter
                (incoming.getOutputStream(), true);
            out.println( "Hello!Enter BYE to exit." );
            boolean done=false;
            while (!done)
            {   String str=in.readLine();
                if (str==null) done=true;
                else
                {   out.println("Echo ("+counter+"): "+str);
                    if (str.trim().equals("BYE"))
                        done=true;
                }
            }
            incoming.close();
        }catch (Exception e)
        {   System.out.println(e);
        }
    }

    private Socket incoming;
    private int counter;
}
```

运行上述程序,它将等待客户的连接。
客户端的程序如下:

```java
import java.net.*;
```

```java
import java.io.*;
import java.awt.*;
import java.awt.event.*;

public class client extends Frame implements ActionListener
{   int i=1;
    Frame f;
    TextField ip, port;
    Label Lip, Lport;
    Button connect, exit;
    public static void main(String args[])
    {   client cl=new client();
        cl.init();
    }

    public void init()
    {   f=new Frame("client connection");
        f.setLayout(new FlowLayout());

        ip=new TextField("127.0.0.1", 20);
        port=new TextField("8189", 5);
        Lip=new Label("ip address");
        Lport=new Label("port");
        connect=new Button("connect");
        exit=new Button("exit");
        connect.addActionListener(this);
        exit.addActionListener(this);
        f.add(Lip);
        f.add(ip);
        f.add(Lport);
        f.add(port);
        f.add(connect);
        f.add(exit);
        f.setSize(500, 60);
        f.setVisible(true);
    }

    public void actionPerformed(ActionEvent e)
    {   if(e.getSource()==exit)
            System.exit(0);
        if(e.getSource()==connect)
        {   new Thread(new threadclient(
                ip.getText(),port.getText(),i)).start();
            i++;
```

```
            }
        }
}

class threadclient extends Frame implements Runnable,ActionListener
{    String ip,port;
    int no;
    Frame f;
    TextArea ta;
    TextField tf;
    Button send,bye;
    InputStream ins;
    Socket s;
    PrintStream out1;
    BufferedReader in;
    BufferedWriter out;
    threadclient(String n,String m,int i)
    {   ip=n;
        port=m;
        no=i;
    }

    public void run()
    {    f=new Frame("clinet NO."+no);
        f.setLayout(new FlowLayout());
        ta=new TextArea("",10,30,TextArea.SCROLLBARS_BOTH);
        tf=new TextField("",20);
        send=new Button("send");
        bye=new Button("bye");
        send.addActionListener(this);
        bye.addActionListener(this);
        f.add(ta);
        f.add(tf);
        f.add(send);
        f.add(bye);
        f.setSize(300, 300);
        f.setVisible(true);
        Integer tmp=new Integer(port);
        int portint=tmp.intValue();
        try
        {    s=new Socket(ip,portint);
            in=new BufferedReader(new InputStreamReader
                (s.getInputStream()));
            out1=new PrintStream(s.getOutputStream());
```

```
            ta.append(in.readLine());
            out1.println("hehe haha");
            ta.append(in.readLine()+"\n");
        }catch(Exception e)
        {   e.printStackTrace();
        }
    }

    public void send(String txt)
    {   try
        {   out1.println(txt);
            out1.flush();
            ta.append(in.readLine()+"\n");
        }catch(IOException e)
        {   e.printStackTrace();
        }
    }

    public void actionPerformed(ActionEvent e)
    {   if(e.getSource()==bye)
        {   send("BYE");
            System.exit(0);
        }
        if(e.getSource()==send)
            send(tf.getText());
    }
}
```

运行客户端程序,将弹出如图 13-1 所示的窗口。

单击窗口中的 connect 按钮,将出现如图 13-2 所示的窗口,同时,服务器端将显示 "Spawning 1"的信息,表示有一个客户已经成功连接。

图 13-1 客户连接窗口

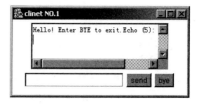

图 13-2 客户端窗口

读者可以通过 connect 按钮,创建多个连接。在资源能够满足的前提下,服务器将会记录下用户创建的所有连接。

在客户端窗口中的最下一个输入框内,读者可以输入一些信息,并单击 send 按钮,这些信息通过服务器发送给当前客户端,并显示在上面的窗口中。单击 bye 按钮后,退出客户程序。

第 2 部分

上 机 指 导

第14章 实验1 熟悉系统及环境

14.1 实验目的

本实验是其他后续实验的基础,要求学生能够正确安装系统并能正确设置环境变量。在此基础上,能够熟悉编写、执行一个应用程序的流程,包括正确使用 javac 和 java 命令。具体目的有以下三个:

(1) 了解 JDK 及 API 文档的安装,能够查阅 API 文档。

(2) 了解 Java 应用程序的调试过程。

(3) 正确使用 Java 命令行的命令。

14.2 实验内容

本实验有两项内容:

1. 安装/熟悉系统

要求在自己使用的机器上安装 JDK 及 API 文档,并正确设置系统环境变量。

2. 编写应用程序 Application,在屏幕上显示如下信息:

```
*******************************
    Welcome  <你的姓名>
*******************************
```

14.3 参考答案

建立一个.java 程序,程序名为 OutputNameApp.java。程序代码如下:

```java
//程序 一个基本的 Java 应用程序
//OutputNameApp.java
public class OutputNameApp
{
    public static void main (String args[])
    {
        System.out.println ("*******************************");
        System.out.println ("      Welcome Name");
        System.out.println ("*******************************");
    }
}
```

14.4 程序说明

使用自己的名字替换程序中的"Name"。然后使用 javac 命令进行编译,编译命令如下:

```
javac OutputNameApp.java
```

如果编译正确,再使用 java 命令执行,执行命令如下:

```
java OutputNameApp
```

此时,就可以在屏幕上看到输出的内容了。

第 15 章 实验 2 简单的输入/输出处理

15.1 实验目的

本实验要求学生能够编写简单的数据处理程序,包括输入/输出处理、数据计算等。输入/输出是每个程序必需的操作,Java 中使用数据流可以实现复杂的输入/输出功能。本实验只编写简单的输入/输出程序,具体的目的是:

(1) 熟悉和理解 Java 中的数据类型、运算符、表达式、程序结构等内容。
(2) 掌握常用的输入/输出方法,编写、调试简单的程序。
(3) 学会使用 Java 提供的类及方法。

15.2 实验内容

本实验完成三个基本的数据处理和输入/输出程序,实验内容包括:

(1) 编写程序,读入用户输入的两个整数 I1 和 I2,然后输出 I1~I2 之间所有为 I1 整数倍数的整数,输出时每行显示的整数个数也由用户来指定。
(2) 编写程序,显示 2~100 中的素数,每行显示 5 个数。
(3) 编写程序,显示当月日历,一个星期占一行,例如:

2016 年 12 月

周日	周一	周二	周三	周四	周五	周六
				1	2	3
4	5	6	7	8	9	10
11	12	13	14	15	16	17
18	19	20	21	22	23	24
25	26	27	28	29	30	31

15.3 参考答案

1. 这是一个应用程序,练习使用 Java 提供的 API 方法。程序代码如下:

```
//程序 一个简单的 Java 应用程序
//Multiples.java
import java.util.Scanner;
public class Multiples
{
    public static void main (String[] args)
    {
        int value, limit, perline, mult, count=0;
```

```java
        Scanner scan=new Scanner (System.in);

        System.out.print ("输入第一个正整数,既是起始数也是间隔数 I1: ");
        value=scan.nextInt();

        System.out.print ("输入上界值: ");
        limit=scan.nextInt();

        System.out.print ("每行输出的数据个数: ");
        perline=scan.nextInt();

        System.out.println ();
        System.out.println ("在 "+value+" 和 "+limit+" (含)之间, "
            +value +" 的倍数有:");

        for (mult=value; mult <=limit; mult +=value)
        {
            System.out.print (mult+"\t");
            //每行输出的数据个数由用户来指定
            count++;
            if (count%perline==0)
                System.out.println();
        }
    }
}
```

2. 程序代码如下:

```java
//程序 一个简单的Java应用程序
//PrimeNumber.java
import java.io.*;

public class PrimeNumber
{   private int pm;
    public void setPm(int pm)
    {   this.pm=pm;
    }
    public boolean isPrime()                    //判断素数
    {   boolean bl=true;
        int i=2;
        for(i=2; i<=Math.sqrt(pm);)             //循环判定有否因子
        {   if(pm%i==0)
            {   bl=false;
                break;                          //如果存在因子,则跳出循环
            }
```

```
            else   i++;                          //否则继续
        }
        return bl;
    }

    public static void main(String args[])
    {   PrimeNumber prim=new PrimeNumber();
        int count=0, i;
        for (i=2; i<=100; i++)
        {   prim.setPm(i);
            boolean bl=prim.isPrime();
            if(bl==true)
            {   System.out.print(i+"  ");
                count++;
                if (count%5==0)
                    System.out.println();
            }
        }
    }
}
```

3. 程序代码如下：

```
//显示当前月份的日历
//MyCalender.java
import java.util.*;

public class MyCalender {
    public static void main(String args[]) {
        Calendar cal=Calendar.getInstance();
        int   day,month,year,date;

        month=cal.get(Calendar.MONTH)+1;
        year=cal.get(Calendar.YEAR);
        System.out.println(year+"年"+month+"月");

        cal.set(Calendar.DAY_OF_MONTH,1);
        date=cal.get(Calendar.DAY_OF_WEEK);                    //本月1号的位置

        System.out.println("  周日  周一  周二  周三  周四  周五  周六");
        for (int i=1;i<date;i++) System.out.print("      ");
        while (cal.get(Calendar.MONTH)==month-1) {             //输出各日
            System.out.print("  ");
            if ((cal.get(Calendar.DAY_OF_MONTH)<10))
                System.out.print("  "+cal.get(Calendar.DAY_OF_MONTH)+"  ");
```

```
        else    System.out.print("  "+cal.get(Calendar.DAY_OF_MONTH)+" ");

        if (cal.get(Calendar.DAY_OF_WEEK)==7)
            System.out.println();
        day=cal.get(Calendar.DAY_OF_MONTH);
        day=day +1;
        cal.set(Calendar.DAY_OF_MONTH,day);
    }
    System.out.println();
  }
}
```

15.4 程序说明

第 1 个实验中,使用了一个简单文本扫描器 Scanner。这个类的使用比较简单,例如可以使用下面的代码从 System.in 中读取一个整数:

```
Scanner sc=new Scanner(System.in);
int i=sc.nextInt();
```

第 2 个实验中涉及了素数的概念,素数也称为质数,这是数论中很重要的概念。

判别一个整数 N 是否为素数有很多的方法,其中最简单的一个方法是,依次判别从 2 到 sqrt(N)这个区间内的每个整数,看看是否是 N 的因子。如果区间内所有的整数都不是 N 的因子,可以断定 N 为素数,否则 N 不是素数。这个方法是从筛法演变而来的。

第 3 个实验中涉及了日期类型。Java 提供了一个日期类 Calendar,使用其中的 API 方法可以对日期进行各类操作。Calendar 类是一个抽象类,该类提供了一个类方法 getInstance(),以获得此类型的一个通用的对象。

类中提供了常见的用于日期及时间的字段,例如 YEAR、MONTH、DAY_OF_MONTH、HOUR 等。可以通过调用 get()、getTimeInMillis()、getTime()等来获取日期及相应的字段,使用 set()方法来设置日历字段值,此外,更改日历字段的方法有三种,分别是:set()、add()和 roll()。

Calendar 有两种解释日历字段的模式,即 lenient 和 non-lenient。当 Calendar 处于 lenient 模式时,它可接受比它所生成的日历字段范围更大范围内的值。但当 Calendar 重新计算日历字段值,并由 get()返回这些值时,所有日历字段都被标准化。例如,lenient 模式下的 GregorianCalendar 将 MONTH=JANUARY、DAY_OF_MONTH=32 解释为 February 1。

第16章 实验3 类的练习

16.1 实验目的

本实验要求学生能够设计程序中使用的类,能够实现类中的相关方法,完成对类中各成员的访问。具体的目的是:

(1) 熟悉和理解 Java 中的类、方法等概念,掌握类的定义和使用,能够在程序中定义自己使用的类,编写构造方法及成员方法。

(2) 能够创建类的实例,掌握对象的声明和不同访问属性的成员访问方式,正确调用类中的方法。

16.2 实验内容

1. 编写一个线段类 MyLine

要求如下:

(1) 主要属性有两个端点:e1、e2,类型为 Point。

(2) 编写构造方法,如 MyLine(Point p1, Point p2)。

(3) 编写成员方法。

例如:

- 检查线段是否位于第一象限 check()。
- 求线段的长度 length()。
- 判断两条直线是否相交(另一线段作为参数)。
- 求一点到该线段(或延长线)的距离。
- 其他方法。

2. 编写测试程序

16.3 参考答案

程序代码如下:

```java
import java.util.*;
import java.awt.geom.*;
import javax.swing.JOptionPane;

class Point {
    private double x, y;
    Point(double x1, double y1)
    {
        x=x1;
        y=y1;
    }
```

```java
    Point()
    {
        this(0, 0);
    }
    Point(Point p)
    {
        this(p.getX(), p.getY());
    }
    void setX(double x1)
    {   x=x1;
    }
    void setY(double y1)
    {   y=y1;
    }
    double getX()
    {   return x;
    }
    double getY()
    {   return y;
    }
    void moveTo(double x1, int y1) {
        x=x1;
        y=y1;
    }
}

public class MyLine {
    private Point e1, e2;                          //线段的两个端点
    MyLine()                                       //构造方法1,两个缺省端点
    {
        e1=new Point(0.0, 0.0);
        e2=new Point(0.0, 0.0);
    }
    MyLine(Point p1, Point p2)                     //构造方法2,两个端点
    {
        e1=new Point(p1);
        e2=new Point(p2);
    }
    MyLine(Point p1)                               //构造方法3,从原点出发的线段
    {
        e1=new Point(0, 0);
        e2=new Point(p1);
    }
```

```
void setE1(Point p)
{   e1=p;
}
void setE2(Point p)
{   e2=p;
}
Point getE1()
{   return e1;
}
Point getE2()
{   return e2;
}
boolean check()                         //检查线段是否在第一象限
{
    //if (this.e1.getX() >0 && this.e1.getY()>0
    //    && this.e2.getX()>0 && this.e2.getY()>0) return true;
    //this 可以省略
    if(e1.getX() >0 && e1.getY()>0 && e2.getX()>0 && e2.getY()>0) return true;
    else return false;
}
double length()                         //求线段的长度
{
    double leng;
    //leng=Math.sqrt((this.e1.getX()-this.e2.getX()) *
    //   (this.e1.getX()-this.e2.getX())+
    //   (this.e1.getY()-this.e2.getY()) *
    //   (this.e1.getY()-this.e2.getY()));
    //this 可以省略
    leng=Math.sqrt((e1.getX()-e2.getX()) * (e1.getX()-e2.getX())+
        (e1.getY()-e2.getY()) * (e1.getY()-e2.getY()));
    return leng;
}
boolean intersect(MyLine line2)         //判断两条线段是否交叉
{
    boolean result;
    //调用 Line2D 中的方法进行判断
    result=Line2D.linesIntersect(getE1().getX(),getE1().getY(),getE2().getX(),
        getE2().getY(), line2.getE1().getX(), line2.getE1().getY(),
        line2.getE2().getX(), line2.getE2().getY());
    return result;
}
double distance(Point p)                //求点到线段的距离
```

```
{
    double result;
    //调用Line2D中的方法求解
    result=Line2D.ptLineDist(getE1().getX(),getE1().getY(),
            getE2().getX(),getE2().getY(),p.getX(),p.getY());
    return result;
}
public static void main(String args[]) {
    boolean result;
    MyLine line1,line2;

    String s="请在下面对话框中输入第一条直线的两点坐标";
    String s1=JOptionPane.showInputDialog(s+"\n 第一个点的横坐标 x");
    String s2=JOptionPane.showInputDialog(s+"\n 第一个点的纵坐标 y");
    String s3=JOptionPane.showInputDialog(s+"\n 第二个点的横坐标 x");
    String s4=JOptionPane.showInputDialog(s+"\n 第二个点的纵坐标 y");

    Point pone=new Point(Double.parseDouble(s1), Double.parseDouble(s2));
    Point ptwo=new Point(Double.parseDouble(s3), Double.parseDouble(s4));

    s="请在下面对话框中输入第二条直线的两点坐标";
    s1=JOptionPane.showInputDialog(s+"\n 第一个点的横坐标 x");
    s2=JOptionPane.showInputDialog(s+"\n 第一个点的纵坐标 y");
    s3=JOptionPane.showInputDialog(s+"\n 第二个点的横坐标 x");
    s4=JOptionPane.showInputDialog(s+"\n 第二个点的纵坐标 y");

    Point pthree=new Point(Double.parseDouble(s1), Double.parseDouble(s2));
    Point pfour=new Point(Double.parseDouble(s3), Double.parseDouble(s4));

    s1=JOptionPane.showInputDialog("请在下面对话框中输入一个点的横坐标 x");
    s2=JOptionPane.showInputDialog("请在下面对话框中输入一个点的纵坐标 y");
    Point pfive=new Point(Double.parseDouble(s1), Double.parseDouble(s2));

    line1=new MyLine(pone, ptwo);
    line2=new MyLine(pthree, pfour);

    //*****************************************象限
    result=line1.check();
    System.out.print("线段 (("+pone.getX()+","+pone.getY()+")  ("+ptwo.
    getX()+","+ptwo.getY()+"))");
    if (result) System.out.println(" 在第一象限");
    else System.out.println(" 不在第一象限");
```

```
//*****************************************长度
System.out.print("线段 (("+pone.getX()+","+pone.getY()+") ("+ptwo.
    getX()+","+ptwo.getY()+"))");
System.out.println(" 的长度是  "+line1.length());

//*****************************************相交
result=line1.intersect(line2);
System.out.print("线段 (("+pone.getX()+","+pone.getY()+") ("+ptwo.
    getX()+","+ptwo.getY()+"))");
System.out.print(" 与线段 (("+pthree.getX()+","+pthree.getY()+") ("+
    pfour.getX()+","+pfour.getY()+"))");
if (result) System.out.println(" 相交");
else System.out.println(" 不相交");

//*****************************************点到线的距离

System.out.print("点 ("+pfive.getX()+","+pfive.getY()+") 到线段 (("
    +pone.getX()+","+pone.getY()+") ("+ptwo.getX()+","+ptwo.getY()+
    ")) 或其延长线的距离是  "+line1.distance(pfive));

System.exit(0);
    }
}
```

16.4 程序说明

程序中定义了二维点类 Point,并定义了三个构造方法,可以根据不同的参数来创建点。Point 类中还有相应的方法可以获取、设置点的横、纵坐标。在 Point 类的基础上定义了线段类 MyLine,三个构造方法可以根据所给出的不同端点情况创建线段。

Java 提供了一个线段类 Line2D,这是二维线段的抽象类。

要检查线段是否位于第一象限,只需要判断线段的两个端点的横、纵坐标值就可以实现。第一象限中所有点的横、纵坐标都大于零。

知道了线段的两个端点坐标,根据计算公式可以很方便地得到线段的长度。设线段的两端点分别是(x_1, y_1)和(x_2, y_2),则线段的长度由以下公式得到:

$$线段长度 = \sqrt{(x_1 - x_2)^2 + (y_1 - y_2)^2}$$

Line2D 中提供了一个方法 intersectsLine(),可以判断参数所指的线段与实例本身所代表的线段是否相交。类似的方法还有一个,即方法 linesIntersect()。参数表中的 8 个双精度值,分别代表 4 个二维点的横、纵坐标,它们又构成两个线段,该方法即判断这两个线段是否交叉。也可以根据线段的坐标值及相应的数学知识,自己编写程序来判断两线段是否相交。

Line2D 中的方法 ptLineDist()可以返回从点到此线段的距离。这个方法重载了三个不同的方法,可以适用所给的端点及线段的不同形式。也可以根据线段、点的坐标值及相应的数学知识,自己编写程序来求解点到线段的距离。

测试程序中,使用了 Java 提供的 JOptionPane 类,这个类可以弹出标准对话框,用来显示一些提示信息或要求用户输入某些值。方法 showInputDialog()可以完成这个功能。它既可以显示提示内容,也可以要求用户输入所需要的内容。

第17章 实验4 模拟彩票开奖游戏

17.1 实验目的

本实验要求学生能够熟练使用 Java 中的各类流程控制语句来设计程序,具体的目的是:

(1) 熟悉程序的流程控制语句的使用。
(2) 熟练使用循环语句实现重复处理。
(3) 熟练使用单分支及多分支语句实现不同情况的处理。

17.2 实验内容

编写一个彩票开奖的模拟程序。游戏共有两种玩法,一种是 21 选 5 的游戏,即要求玩家输入 5 个 1~21 之间不重复的整数。另一种是 6+1 玩法,即要求玩家输入 7 个整数,代表所购买的彩票的号码,最后一个是特别号。

由用户选择玩法,根据游戏种类,计算机使用随机数发生器在内部生成不同个数不同范围的数据,这些数据最初时是不公开显示给玩家的。计算机接收玩家输入的整数,按照两者相重合数据的个数多少来确定玩家是否获胜及获胜的等级。重合的数据越多,获胜的等级越高。计算机将所生成的随机数与用户输入的号码进行比较,判断是否中奖。如果中奖,还要显示所中的奖项。最后,将玩家输入的数据及计算机内部生成的数据同时显示给玩家。

游戏中奖规则如下:

1. 21 选 5 游戏
- 一等奖:5 个号码相同。
- 二等奖:4 个号码相同。
- 三等奖:3 个号码相同。

2. 6+1 游戏
- 特等奖:全部 7 位数字相同。
- 一等奖:连续 6 位数字相同。
- 二等奖:连续 5 位数字相同。
- 三等奖:连续 4 位数字相同。
- 四等奖:连续 3 位数字相同。

17.3 参考答案

程序代码如下:

```
import java.io.*;
import java.util.StringTokenizer;
import java.util.Random;
```

```java
public class Lottery
{
    public static void main(String[]args)
    {
        int playtimes=0;
        try
        {
            if(args.length==1) playtimes=Integer.parseInt(args[0]);
            else playtimes=1;
        }catch(NumberFormatException e)
        {
            System.out.println(e);
        }
        while(playtimes>0)
        {
            int choice=init();
            switch(choice)
            {
                case 1:
                    int[]numbers_1=new int[5];
                    input_1(numbers_1);
                    check_1(numbers_1);
                    break;
                case 2:
                    int[]numbers_2=new int[7];
                    input_2(numbers_2);
                    check_2(numbers_2);
            }
            playtimes--;
        }
    }
    static int init()
    {
        boolean InputLoopflag=true;
        int n=0;
        System.out.println("请按数字键1或2选择一种玩法(其中1代表"21选5",
        2代表"6+1"):");
        InputStreamReader ir;
        BufferedReader in;
        String s=new String();
        try{
            ir=new InputStreamReader(System.in);
            in=new BufferedReader(ir);
            while(InputLoopflag)
```

```java
        {
            s=in.readLine();
            try{
                n=Integer.parseInt(s);
                if(n!=1&&n!=2)
                {
                    System.out.println("输入错误!请重试:");
                    continue;
                }
                InputLoopflag=false;
            }catch(NumberFormatException e)
            {
                System.out.println("非数字!请重试:");
                continue;
            }
        }
    } catch (IOException e) {   System.out.println(e);}
    return n;
}
static void input_1(int[]numbers_1)
{
    System.out.println("请输入"+numbers_1.length+"个数(这"
    +numbers_1.length+"个数必须互不相同,且在1~21之间):");
    boolean InputLoopflag=true;
    StringTokenizer st;
    InputStreamReader ir;
    BufferedReader in;
    String s=new String();
    try{
        ir=new InputStreamReader(System.in);
        in=new BufferedReader(ir);
        L1:
        while(InputLoopflag)
        {
            s=in.readLine();
            st=new StringTokenizer(s);
            int count=st.countTokens();
            if(count!=numbers_1.length)
            {
                System.out.println("只能输"+numbers_1.length+"个!请重试:");
                continue;
            }
            try{
                for(int i=0;i<numbers_1.length;i++)
```

```java
                {
                    numbers_1[i]=Integer.parseInt(st.nextToken());
                    for(int j=0;j<i;j++)
                        if(numbers_1[i]==numbers_1[j])
                        {
                            System.out.println(numbers_1.length+
                                "个数必须互不相同!请重试:");
                            continue L1;
                        }
                }
                for(int i=0;i<numbers_1.length;i++)
                    if(numbers_1[i]<1||numbers_1[i]>21)
                    {
                        System.out.println(numbers_1.length+"个数
                            必须在1~21之间!请重试:");
                        continue L1;
                    }
            }catch(NumberFormatException e)
            {
                System.out.println("非数字!请重试:");
                continue;
            }
            InputLoopflag=false;
        }
    }catch (IOException e)
    {
        System.out.println(e);
    }
}
static void input_2(int[]numbers_2)
{
    System.out.println("请连续输入"+numbers_2.length+"位数字
        (代表购买彩票的号码,最后一位为特别号):");
    boolean b=true;
    InputStreamReader ir;
    BufferedReader in;
    String s=new String();
    try{
        ir=new InputStreamReader(System.in);
        in=new BufferedReader(ir);
        L2:
        while(b)
        {
            s=in.readLine();
```

```
            if(s.length()!=numbers_2.length)
            {
                System.out.println("只能输"+numbers_2.length+"位！
                请重试:");
                continue;
            }
            for(int i=0;i<numbers_2.length;i++)
                numbers_2[i]=s.charAt(i)-48;
            for(int i=0;i<numbers_2.length;i++)
                if(numbers_2[i]<0||numbers_2[i]>9)
                {
                    System.out.println("只能输入数字!请重试:");
                    continue L2;
                }
            b=false;
        }
    } catch (IOException e) {  System.out.println(e);}
}
static void check_1(int[]numbers)
{
    System.out.println("\n21选5的规则是:\n一等奖:5个号码相同;\n二等奖:
    4个号码相同;\n三等奖:3个号码相同。\n");
    int[]award=new int[5];
    Random ran=new Random();
    L3:
    for(int i=0;i<award.length;i++)
    {
        award[i]=Math.abs(ran.nextInt())%21+1;
        for(int j=0;j<i;j++)
            if(award[j]==award[i])
            {
                i--;
                continue L3;
            }
    }
    int same=0;
    L4:
    for(int i=0;i<numbers.length;i++)
        for(int j=0;j<award.length;j++)
            if(numbers[i]==award[j])
            {
                same++;
                continue L4;
            }
```

```java
        switch(same)
        {
            case 5:
                System.out.println("恭喜你!你中了一等奖,奖金 5000000!!!");
                break;
            case 4:
                System.out.println("恭喜你!你中了二等奖,奖金 50000!!!");
                break;
            case 3:
                System.out.println("恭喜你!你中了三等奖,奖金 5000!!!");
                break;
            default:
                System.out.println("很遗憾,你没有中奖,祝你下次好运。");
        }
        System.out.print("随机产生的中奖号码是:");
        for(int i=0;i<award.length;i++) System.out.print(award[i]+"  ");
        System.out.print("\n 你选的号码是       :");
        for(int i=0;i<numbers.length;i++) System.out.print(numbers[i]+"  ");
        System.out.println("\n");
    }
    static void check_2(int[]numbers)
    {
        System.out.println("\n6+1 的规则是:");
        System.out.println("特等奖:6+1 位数都相同;\n"+
            "一等奖:连续 6 位数相同;\n"+
            "二等奖:连续 5 位数相同;\n"+
            "三等奖:连续 4 位数相同;\n"+
            "四等奖:连续 3 位数相同。\n");

        int[]award=new int[7];
        int i;
        Random ran=new Random();
        for(i=0;i<award.length;i++) award[i]=Math.abs(ran.nextInt())%10;

        for(i=0;i<award.length-1;i++)
            if(award[i]!=numbers[i]) break;

        if(i==6)
        {
            if(award[i]==numbers[i])
                System.out.println("恭喜你!你中了特等奖,奖金 5000000!!!");
            else System.out.println("恭喜你!你中了一等奖,奖金 500000!!!");
            System.out.print("随机产生的中奖号码是:");
            for(i=0;i<award.length;i++) System.out.print(award[i]);
```

```java
        System.out.print("\n你选的号码是        :");
        for(i=0;i<numbers.length;i++) System.out.print(numbers[i]);
        return;
    }
    int same1=0,same2=0;
    for(i=0;i<award.length-1;i++)
    {
        if(award[i]==numbers[i]) same1++;
        else
        {
            if(same1>same2) same2=same1;
            same1=0;
        }
    }
    int same=same1>same2?same1:same2;
    switch(same)
    {
        case 5:
            System.out.println("恭喜你!你中了二等奖,奖金 50000!!!");
            break;
        case 4:
            System.out.println("恭喜你!你中了三等奖,奖金 5000!!!");
            break;
        case 3:
            System.out.println("恭喜你!你中了三等奖,奖金 500!!!");
            break;
        default:
            System.out.println("很遗憾,你没有中奖,祝你下次好运。");
    }
    System.out.print("随机产生的中奖号码是:");
    for(i=0;i<award.length;i++) System.out.print(award[i]);
    System.out.print("\n你选的号码是        :");
    for(i=0;i<numbers.length;i++) System.out.print(numbers[i]);
    System.out.println("\n");
    }
}
```

17.4 程序说明

本实验中,使用了另一类输入/输出处理,即使用数据流完成输入/输出功能。可以使用如下语句接收数据:

```
//从标准输入中读入数据,存放到已定义的变量中
ir=new InputStreamReader(System.in);
in=new BufferedReader(ir);
s=in.readLine();
```

输入的内容必须与接收数据的变量的类型一致。如果类型不一致,可以使用强制类型转换。例如,输入的内容是字符串形式,而需要保存到整型变量中,则将输入的内容转变为整数,使用 Integer 类中的 parseInt()方法,如下所示:

```
//将字符串 s 转换为整数类型
n=Integer.parseInt(s);
```

程序中还用到了 Integer 类,它是 Java 中的包装类(也叫封装类,wrapper class)。Java 是面向对象的语言,它提供的方法只能用于对象,而不能用于基本类型。即不能对原始类型调用方法,但可以对对象使用这些方法。当想对基本数据施加某个方法时,必须使用包装类将基本类型转化为对象,这样,程序中就可以将基本数据视作对象并连接相关的方法。

Java 为每个原始类型提供了包装类,这些包装类有:Boolean、Byte、Short、Character、Integer、Long、Float 和 Void 等。除首字母大写外,包装类与基本类型的名字是一样的。

程序运行时,要求输入多个数据,并且在一行内输入完毕,这里使用了 StringTokenizer 类。它是 java.lang.Object 的子类。

StringTokenizer 的作用是每次返回字符串内的一个记号,即它的一个子串。至于子串中的符号是什么,则不去区分。子串之间的分隔符可以是制表符、空格以及换行符。例如,当用空格当作分隔符时,字符串"Where is my cat?"中的各记号分别是"Where""is""my"和"cat?"。当然也可以不使用这些预定的分隔符。当使用特殊的分隔符时,需要在构造方法中指定。

程序中还使用了伪随机数类 Random。它通过一个种子及特殊的计算公式,得到随机数序列。Random 类中提供了多个功能类似的方法,可以得到不同类型的伪随机数。例如,调用 nextFloat()可以得到 0~1 之间的随机浮点数,调用 nextDouble()可以得到 0~1 之间的随机双精度数,调用 nextInt()可以得到随机整数。

第18章 实验5 模拟CD出租销售店

18.1 实验目的

本实验要求学生能够熟练使用Java中的各类流程控制语句来设计程序,掌握基本的程序设计方法,具体的目的是:

(1) 理解解决问题的技巧,按照自顶向下、逐步求精的过程设计程序。
(2) 学习设计算法的初步知识。
(3) 学习异常的处理方法。
(4) 初步了解文件的处理方法。

18.2 实验内容

假设你在业余时间经营一个会员制的CD出租销售店,需要一个管理程序。管理程序的基本功能有两方面:一是对会员的管理,包括增加会员、删除会员;二是对货品的管理,包括出租、销售CD、进货、统计账目等。要求设计程序实现这些功能。

18.3 参考答案

程序代码如下:

```java
import java.io.*;
class MyCD                          //CD类
{
    private long serialnumber;      //编号
    private String name;            //CD名称
    private float price;            //购入价格
    private float sailprice;        //销售价格
    private float lendprice;        //出租价格
    private int count;              //已出租的次数
    private int state;              //CD状态:-1表示已出租,0表示在库中,1表示已销售
    MyCD(long k,String n,float p,float sp,float lp)
    {
        serialnumber=k;
        name=n;
        price=p;
        lendprice=lp;
        sailprice=sp;
        count=0;
        state=0;
    }
    long getSerial()
```

```
    {
        return serialnumber;
    }
    String getName()
    {
        return name;
    }
    float getPrice()
    {
        return price;
    }
    float getLendPrice()
    {
        return lendprice;
    }
    float getSailPrice()
    {
        return sailprice;
    }
    int getCount()
    {
        return count;
    }
    int getState()
    {
        return state;
    }
    void setSerial(long k)
    {
        serialnumber=k;
    }
    void setName(String n)
    {
        name=n;
    }
    void setLendPrice(float lp)
    {
        lendprice=lp;
    }
    void setSailPrice(float sp)
    {
        sailprice=sp;
    }
    void lend()                    //改变CD为出租状态
```

```java
    {
        state=-1;
        count++;
    }
    void returned()                                    //CD已归还
    {
        state=0;
    }
    void sail()                                        //CD已销售
    {
        state=1;
    }
    void output(String text, boolean app)              //输出CD信息
    {
        try{
            DataOutputStream oos=new DataOutputStream(new BufferedOutputStream
                (new FileOutputStream(text,app)));
            oos.writeLong(serialnumber);
            oos.writeUTF(name);
            oos.writeFloat(price);
            oos.writeFloat(lendprice);
            oos.writeFloat(sailprice);
            oos.writeInt(count);
            oos.writeInt(state);
            oos.close();
        }catch(IOException e)
        {
            System.out.println(e);
        }
    }
    boolean input(DataInputStream ois)                 //读入CD信息
    {
        try{
            serialnumber=ois.readLong();
            name=ois.readUTF();
            price=ois.readFloat();
            lendprice=ois.readFloat();
            sailprice=ois.readFloat();
            count=ois.readInt();
            state=ois.readInt();
        }catch(IOException e)
        {
            return false;
        }
```

```
            return true;
        }
    }
    class Member                                      //会员类的定义
    {
        private String name;                          //会员名
        private String keynumber;                     //密码
        private boolean hasborrowed;                  //是否有未归还的 CD?
        private long whichCD;                         //借的是哪张 CD?
        Member(String n,String k)
        {
            name=n;
            keynumber=k;
            hasborrowed=false;
            whichCD=0L;
        }
        String getName()
        {
            return name;
        }
        String getKeyNumber()
        {
            return keynumber;
        }
        boolean getHasBorrowed()
        {
            return hasborrowed;
        }
        long getWhichCD()
        {
            return whichCD;
        }
        void setName(String s)
        {
            name=s;
        }
        void change(long w)
        {
            hasborrowed=(hasborrowed==false?true:false);
            whichCD=w;
        }
        void output(String text, boolean app)         //输出会员信息
        {
            try{
```

```
        DataOutputStream oos=new DataOutputStream(new BufferedOutputStream
            (new FileOutputStream(text,app)));
        oos.writeUTF(name);
        oos.writeUTF(keynumber);
        oos.writeBoolean(hasborrowed);
        oos.writeLong(whichCD);
        oos.close();
        }catch(IOException e){ System.out.println(e);}
    }
    boolean input(DataInputStream ois)                //读入会员信息
    {
        try{
            name=ois.readUTF();
            keynumber=ois.readUTF();
            hasborrowed=ois.readBoolean();
            whichCD=ois.readLong();
        }catch(IOException e)
        {
            return false;
        }
        return true;
    }
}
public class CDMain
{
    public static void main(String[]args)
    {
        initial();              //首次运行程序时需要先初始化,以后的执行要跳过
        int choice=0;
        boolean continu_e=true;
        InputStreamReader ir;
        BufferedReader in;
        String s=new String();
        float benefits;
        while(continu_e)
        {
            System.out.println("请选择你要进行的操作(请输入1～8中的任一数字):\n"
                +"1:增加会员;\n"
                +"2:删除会员;\n"
                +"3:出租 CD;\n"
                +"4:销售 CD;\n"
                +"5:进货;\n"
                +"6:统计收支情况;\n"
                +"7:归还 CD;\n"
```

```java
            +"8:退出本系统。\n");
        boolean b=true;
        try{
            ir=new InputStreamReader(System.in);
            in=new BufferedReader(ir);           //输入要进行的操作所对应的编号
            while(b)
            {
                s=in.readLine();
                try{
                    choice=Integer.parseInt(s);
                    if(choice<1||choice>8)       //编号超出范围
                    {
                        System.out.println("输入错误!请重试:");
                        continue;
                    }
                    b=false;
                }catch(NumberFormatException e)
                {
                    System.out.println("非数字!请重试:");
                    continue;
                }
            }
        } catch (IOException e)
        {
            System.out.println(e);
        }
        switch(choice)              //根据编号调用相应的方法实现相应的功能
        {
        case 1:                     //增加会员
            addMember();
            break;
        case 2:                     //删除会员
            deleteMember();
            break;
        case 3:                     //出租CD
            rentOrSail(true);
            break;
        case 4:                     //销售CD
            rentOrSail(false);
            break;
        case 5:                     //进货,即添加CD
            addCD();
            break;
        case 6:                     //统计收支情况
```

```
                benefits=compute();
                break;
            case 7:                                   //归还CD
                returnto();
                break;
            default:                                  //退出系统
                continu_e=false;
            }
        }
        System.out.println("谢谢您的使用,再见!\n");
        System.exit(0);
    }
    static void initial()                             //初始化,首次运行程序时执行
    {
        MyCD CD[]=new MyCD[10];                       //保存CD信息,最多10张
        CD[0]=new MyCD(1L,"张信哲",5.0f,8.0f,0.5f);
        CD[1]=new MyCD(2L,"罗大佑",7.0f,15.0f,1.5f);
        CD[2]=new MyCD(3L,"任贤齐",10.0f,15.0f,1.0f);
        CD[3]=new MyCD(4L,"羽泉",9.0f,10.0f,1.0f);
        CD[4]=new MyCD(5L,"刘德华",12.0f,15.0f,2.0f);
        CD[5]=new MyCD(6L,"张学友",6.0f,12.0f,1.0f);
        CD[6]=new MyCD(7L,"齐秦",3.0f,5.0f,1.0f);
        CD[7]=new MyCD(8L,"孙楠",15.0f,15.0f,1.5f);
        CD[8]=new MyCD(9L,"阿杜",8.0f,10.0f,1.5f);
        CD[9]=new MyCD(10L,"那英",7.0f,12.0f,1.5f);
        for(int i=0;i<CD.length;i++)
            CD[i].output("MyCD.txt",true);
        Member m[]=new Member[3];                     //保存会员信息,最多3人
        m[0]=new Member("qwe1","qwe123");
        m[1]=new Member("asd2","asd234");
        m[2]=new Member("zxc3","zxc345");
        for(int i=0;i<m.length;i++)
            m[i].output("members.txt",true);
    }
    static void addMember()
    {
        InputStreamReader ir;
        BufferedReader in;
        String na=new String();
        String kn=new String();
        boolean b=true;
        try
        {
            ir=new InputStreamReader(System.in);
```

```java
            in=new BufferedReader(ir);
            System.out.print("请输入新会员的名字:");
            L1:
            while(b)
            {
                na=in.readLine();
                try{
                    DataInputStream ois=new DataInputStream(new
                        BufferedInputStream(new FileInputStream("members.txt")));
                    Member m=new Member(" "," ");
                    while(m.input(ois))
                    {
                        if(na.equals(m.getName()))
                        {
                            System.out.print("这个名字已有人使用,请重新输入:");
                            ois.close();
                            continue L1;
                        }
                    }
                    ois.close();
                } catch(IOException e)
                {
                    System.out.println(e);
                }
                System.out.print("请输入密码:");
                kn=in.readLine();
                b=false;
            }
        }catch(IOException e)
        {
            System.out.println(e);
        }
        Member m1=new Member(na,kn);
        m1.output("members.txt",true);
        System.out.println("已成功增加会员!\n\n");
    }
    static void deleteMember()
    {
        InputStreamReader ir;
        BufferedReader in;
        String na=new String();
        String kn=new String();
        try
        {
```

```
boolean isinlist=false,isfirst=true;
ir=new InputStreamReader(System.in);
in=new BufferedReader(ir);
System.out.print("请输入要删除的会员的名字:");
na=in.readLine();
try{
    DataInputStream ois=new DataInputStream(new BufferedInputStream
        (new FileInputStream("members.txt")));
    Member m=new Member(" "," ");
    while(m.input(ois))
    {
        if(na.equals(m.getName()))
        {
            if(m.getHasBorrowed())
            {
                System.out.println("此人有未归还的 CD,不能删除!\n\n");
                ois.close();
                return;
            }
            else
            {
                isinlist=true;
                kn=m.getKeyNumber();
                continue;
            }
        }
        if(isfirst)
        {
            m.output("temp_members.txt",false);
            isfirst=false;
        }
        else m.output("temp_members.txt",true);
    }
    ois.close();
    isfirst=true;
    ois=new DataInputStream(new BufferedInputStream
        (new FileInputStream("temp_members.txt")));
    while(m.input(ois))
    {
        if(isfirst)
        {
            m.output("members.txt",false);
            isfirst=false;
        }
```

```java
                else m.output("members.txt",true);
            }
        }catch(IOException e)
        {
            System.out.println(e);
        }
        if(isinlist)
            System.out.println("已成功删除会员"+na+",此会员的密码为"+kn+"!\n\n");
        else System.out.println("没有名字为"+na+"的会员!\n\n");
    } catch(IOException e)
    {
        System.out.println(e);
    }
}
static void rentOrSail(boolean ros)
{
    InputStreamReader ir;
    BufferedReader in;
    long sn=0L;
    String na=new String();
    String kn=new String();
    try
    {
        boolean ismember = false, isinlist = false, isfirst = true, b = true,
        exist=false;
        ir=new InputStreamReader(System.in);
        in=new BufferedReader(ir);
        System.out.print("请输入请求操作的会员的名字:");
        na=in.readLine();
        System.out.print("请输入请求操作的会员的密码:");
        kn=in.readLine();
        try{
            DataInputStream ois=new DataInputStream(new BufferedInputStream
                (new FileInputStream("members.txt")));
            Member m=new Member(" "," ");
            while(m.input(ois))
            {
                if(na.equals(m.getName()))
                {
                    ismember=true;
                    if(kn.equals(m.getKeyNumber()))
                    {
                        if(m.getHasBorrowed()&&ros)
                            System.out.println("此会员还有一张CD未还,
```

```java
                    不能再借!\n\n");
                else
                {
                    System.out.println("此会员有权进行此项操作。");
                    isinlist=true;
                }
            }
            else System.out.println("此会员的密码不正确!请退出。\n\n");
            break;
        }
    }
    if(!ismember) System.out.println("没有这个会员!!\n\n");
    ois.close();
    if(isinlist)
    {
        MyCD mCD=new MyCD(0L," ",0,0,0);
        System.out.println("库中的可用CD有:\n");
        System.out.println("编号\t名称\t出租价格\t销售价格\t");
        ois=new DataInputStream(new BufferedInputStream
        (new FileInputStream("MyCD.txt")));
        while(mCD.input(ois))
        {
            if(mCD.getState()==0)
                System.out.println(mCD.getSerial()
                +"\t"+mCD.getName()
                +"\t"+mCD.getLendPrice()
                +"\t"+mCD.getSailPrice());
        }
        ois.close();
        System.out.print("请输入想要的CD的编号:");
        while(b)
        {
            kn=in.readLine();
            try{
                sn=(long)Integer.parseInt(kn);
                if(sn<=0)
                {
                    System.out.print("非正整数!请重试:");
                    continue;
                }
            }catch(NumberFormatException e)
            {   System.out.print("非数字!请重试:");
                continue;
            }
```

```
            b=false;
    }
    ois=new DataInputStream(new BufferedInputStream
        (new FileInputStream("MyCD.txt")));
    while(mCD.input(ois))
    {
        if(sn==mCD.getSerial())
        {
            if(mCD.getState()==0)
            {
                if(ros) mCD.lend();
                else mCD.sail();
                exist=true;
            }
        }
        if(isfirst)
        {
            mCD.output("temp_MyCD.txt",false);
            isfirst=false;
        }
        else mCD.output("temp_MyCD.txt",true);
    }
    isfirst=true;
    ois=new DataInputStream(new BufferedInputStream
        (new FileInputStream("temp_MyCD.txt")));
    while(mCD.input(ois))
    {
        if(isfirst)
        {
            mCD.output("MyCD.txt",false);
            isfirst=false;
        }
        else mCD.output("MyCD.txt",true);
    }
    if(exist)
    {
        isfirst=true;
        ois=new DataInputStream(new BufferedInputStream
            (new FileInputStream("members.txt")));
        while(m.input(ois))
        {
            if(na.equals(m.getName())&&ros) m.change(sn);
            if(isfirst)
            {
```

```java
                    m.output("temp_members.txt",false);
                    isfirst=false;
                }
                else m.output("temp_members.txt",true);
            }
            isfirst=true;
            ois=new DataInputStream(new BufferedInputStream
                (new FileInputStream("temp_members.txt")));
            while(m.input(ois))
            {
                if(isfirst)
                {
                    m.output("members.txt",false);
                    isfirst=false;
                }
                else m.output("members.txt",true);
            }
            if(ros) System.out.println("名字为"+na+"的会员
                已成功租借编号为"+sn+"的CD!\n\n");
            else System.out.println("名字为"+na+"的会员
                已成功购买编号为"+sn+"的CD!\n\n");
        }
        else System.out.println("没有找到编号为"+sn+"的CD!!!\n\n");
    }
    } catch(IOException e)
    {
        System.out.println(e);
    }
} catch(IOException e)
{
    System.out.println(e);
}
}
static void returnto()
{
    InputStreamReader ir;
    BufferedReader in;
    long sn=0L;
    String na=new String();
    try
    {
        boolean ismember=false,isfirst=true;
        ir=new InputStreamReader(System.in);
        in=new BufferedReader(ir);
```

```java
System.out.print("请输入请求操作的会员的名字:");
na=in.readLine();
try{
    DataInputStream ois=new DataInputStream(new BufferedInputStream
        (new FileInputStream("members.txt")));
    Member m=new Member(" "," ");
    while(m.input(ois))
    {
        if(na.equals(m.getName())&&m.getHasBorrowed())
        {
            sn=m.getWhichCD();
            m.change(0L);
            ismember=true;
        }
        if(isfirst)
        {
            m.output("temp_members.txt",false);
            isfirst=false;
        }
        else m.output("temp_members.txt",true);
    }
    ois.close();
    if(!ismember) System.out.println("没有这个会员或此人未借CD!!\n\n");
    else
    {
        isfirst=true;
        ois=new DataInputStream(new BufferedInputStream
            (new FileInputStream("temp_members.txt")));
        while(m.input(ois))
        {
            if(isfirst)
            {
                m.output("members.txt",false);
                isfirst=false;
            }
            else m.output("members.txt",true);
        }
        isfirst=true;
        MyCD mCD=new MyCD(0L," ",0,0,0);
        ois=new DataInputStream(new BufferedInputStream
            (new FileInputStream("MyCD.txt")));
        while(mCD.input(ois))
        {
            if(sn==mCD.getSerial()) mCD.returned();
```

```java
                if(isfirst)
                {
                    mCD.output("temp_MyCD.txt",false);
                    isfirst=false;
                }
                else mCD.output("temp_MyCD.txt",true);
            }
            isfirst=true;
            ois=new DataInputStream(new BufferedInputStream
                (new FileInputStream("temp_MyCD.txt")));
            while(mCD.input(ois))
            {
                if(isfirst)
                {
                    mCD.output("MyCD.txt",false);
                    isfirst=false;
                }
                else mCD.output("MyCD.txt",true);
            }
            System.out.println("名字为"+na+"的会员已把编号
                为"+sn+"的 CD 归还!\n\n");
            }
        } catch(IOException e)
        {
            System.out.println(e);
        }
    } catch(IOException e)
    {
        System.out.println(e);
    }
}
static void addCD()
{
    InputStreamReader ir;
    BufferedReader in;
    long sn=0L;
    String s=new String();
    String na=new String();
    float p=0.0f,lp=0.0f,sp=0.0f;
    boolean b=true;
    try
    {
        ir=new InputStreamReader(System.in);
        in=new BufferedReader(ir);
```

```java
        System.out.print("请输入新CD的编号(正的长整数):");
L1:
    while(b)
    {
        s=in.readLine();
        try{
            sn=(long)Integer.parseInt(s);
            if(sn<=0)
            {
                System.out.print("非正整数!请重试:");
                continue L1;
            }
        }catch(NumberFormatException e)
        {
            System.out.print("非数字!请重试:");
            continue L1;
        }
        try{
            DataInputStream ois=new DataInputStream(new
                BufferedInputStream(new FileInputStream("MyCD.txt")));
            MyCD CD=new MyCD(0L," ",0.0f,0.0f,0.0f);
            while(CD.input(ois))
            {
                if(sn==CD.getSerial())
                {
                    System.out.print("这个编号已使用,请重新输入:");
                    ois.close();
                    continue L1;
                }
            }
            ois.close();
        } catch(IOException e)
        {
            System.out.println(e);
        }
        System.out.print("请输入CD名称:");
        na=in.readLine();
        System.out.print("请输入CD的购买价格(正浮点数):");
        boolean ok=true;
L2:
        while(ok)
        {
            s=in.readLine();
            try{
```

```
                p=Float.parseFloat(s);
                if(p<=0)
                {
                    System.out.print("非正浮点数!请重试:");
                    continue L2;
                }
                ok=false;
            }catch(NumberFormatException e)
            {
                System.out.println("非浮点数!请重试:");
                continue L2;
            }
        }
ok=true;
System.out.print("请输入CD的出租价格(正浮点数):");
L3:
    while(ok)
    {
        s=in.readLine();
        try{
            lp=Float.parseFloat(s);
            if(lp<=0)
            {
                System.out.print("非正浮点数!请重试:");
                continue L3;
            }
            ok=false;
        }catch(NumberFormatException e)
        {
            System.out.println("非浮点数!请重试:");
            continue L3;
        }
    }
ok=true;
System.out.print("请输入CD的销售价格(正浮点数):");
L4:
    while(ok)
    {
        s=in.readLine();
        try{
            sp=Float.parseFloat(s);
            if(sp<=0)
            {
                System.out.print("非正浮点数!请重试:");
```

```
                        continue L4;
                    }
                    ok=false;
                }catch(NumberFormatException e)
                {
                    System.out.println("非浮点数!请重试:");
                    continue L4;
                }
            }
            b=false;
        }
    } catch(IOException e)
    {
        System.out.println(e);
    }
    MyCD mCD=new MyCD(sn,na,p,sp,lp);
    mCD.output("MyCD.txt",true);
    System.out.println("已成功增加 CD!\n\n");
}
static float compute()
{
    float investment=0.0f,abtained=0.0f;
    try{
        DataInputStream ois=new DataInputStream(new BufferedInputStream
            (new FileInputStream("MyCD.txt")));
        MyCD mCD=new MyCD(0L," ",0,0,0);
        while(mCD.input(ois))
        {
            investment +=mCD.getPrice();
            abtained +=mCD.getLendPrice() * mCD.getCount();
            if(mCD.getState()==1) abtained +=mCD.getSailPrice();
        }
        ois.close();
    }catch(IOException e)
    {
        System.out.println(e);
    }
    System.out.println("投入的资金为:"+investment+"元人民币。");
    System.out.println("回收的资金为:"+abtained+"元人民币。");
    System.out.println("收支的差额为:"+ (abtained-investment)+"元人民币。\n\n");
    return (abtained-investment);
}
}
```

18.4 程序说明

本实验使用文件保存会员信息及 CD 信息。会员信息包括会员名和密码,保存在文件 members.txt 中。CD 信息包括 CD 的编号、CD 名称、购入价格、销售价格和出租价格,保存在文件 MyCD.txt 中。

首次执行程序时,需要先调用 initial()方法,建立必要的初始数据。目前假设有 10 张 CD 和 3 名会员。再次执行时,跳过初始化部分。

MyCD 类中,除了对 CD 各属性进行访问的一组方法外,还有输入/输出方法,可以实现从文件中读入信息或向文件中写出信息的功能。Member 类中也有对实例属性进行操作的一组方法,以及将信息写入文件或从文件中读出的输入/输出方法。

编写输入/输出程序时,必须进行异常控制,程序中输入/输出部分都必须含在 try 语句块内。

第19章 实验6 计算器

19.1 实验目的

本实验要求学生能够掌握初步的图形界面设计方法及事件处理机制,实现简单的功能,具体的目的是:

(1) 熟悉和理解 Java 中 AWT 及 Swing,能够设计简单美观的用户界面。
(2) 掌握使用 Java 中图形界面设计的基本元素与方法。

19.2 实验内容

创建一个 Frame,完成一个计算器的基本功能,如图19-1所示。

图 19-1 计算器

19.3 参考答案

程序代码如下:

```java
import javax.swing.*;
import java.awt.*;
import java.awt.event.*;

public class MyComputer extends JFrame implements ActionListener
{
    JTextArea memoryArea=new JTextArea(" ",1, 3);
    JTextArea dispresult=new JTextArea("0.     ",1, 20);

    Font font=new Font("Arial Rounded MT Bold",Font.PLAIN,15);   //显示的字体格式

    JButton clear=new JButton("C");                              //清除键
    JButton[]jbuttons=new JButton[24];                           //各按键

    double result=0, first=0, second=0;
    double memery=0;
    char firstsymbol='\0', secondsymbol='\0';
    boolean prev=true, repeat=true, dot=true;
    MyComputer()                            //定义计算器的显示样式
    {
        super("Calculator");
        try
        {
            UIManager.setLookAndFeel(UIManager.getSystemLookAndFeelClassName());
        }catch(Exception e)
```

```java
        {
            System.out.println(e);
        }
        JPanel resultField=new JPanel();      //用来放置结果、清除键、保存标识的区域,
                                              //上半部分
        JPanel buttonField=new JPanel();      //用来放置计算器上的各按键,下半部分
        Container all=getContentPane();       //组合上述两部分

        GridLayout grid1;
        grid1=new GridLayout(4,6,3,3);        //放置按键

        clear.setFont(font);                  //规定各部分的显示字体样式
        memoryArea.setFont(font);
        dispresult.setFont(font);
        memoryArea.setEditable(false);
        clear.addActionListener(this);
        dispresult.setEditable(false);

        resultField.add(memoryArea);
        resultField.add(clear);
        resultField.add(dispresult);

        all.setLayout(new FlowLayout());
        all.add(resultField);
        String[]buttonname={"sin","MC","0","1","2","+","cos","MR","3","4","5","-","tan","MS","6","7","8","x","+/-","M+","9",".","=","/"};
                                              //各按键显示名称

        buttonField.setLayout(grid1);
        for(int i=0;i<6;i++)
        {
            for(int j=0;j<4;j++)
            {
                jbuttons[i*4+j]=new JButton(buttonname[i*4+j]);
                jbuttons[i*4+j].addActionListener(this);  //为各按键注册侦听器
                jbuttons[i*4+j].setFont(font);
                buttonField.add(jbuttons[i*4+j]);
            }
        }
        all.add(buttonField);
        setSize(400,210);
        setResizable(true);
        setVisible(true);
    }
```

```java
public void pressNumber(String n)
{
    if(prev)
    {
        dispresult.setText(n);
        prev=false;
    }
    else dispresult.append(n);
}
public boolean divide(double d)
{
    if(d==0)
    {
        dispresult.setText("除数不能为 0!");
        prev=true;
        repeat=true;
        firstsymbol='\0';
        secondsymbol='\0';
        return true;
    }
    return false;
}
public void actionPerformed(ActionEvent e)            //对按键事件的响应
{
    Object source=e.getSource();
    //判断事件源,并进行相应的处理
    if(source==clear)
    {
        dispresult.setText("0.");
        firstsymbol='\0';
        secondsymbol='\0';
        prev=true;
        repeat=true;
        dot=true;
        return;
    }
    if(source==jbuttons[0])
    {
        double temp=Math.sin(Double.parseDouble(dispresult.getText()));
        dispresult.setText(String.valueOf(temp));
        prev=true;
        repeat=false;
        dot=true;
        return;
```

```
    }
    if(source==jbuttons[1])
    {
        memery=0;
        memoryArea.setText("   ");
        prev=true;
        repeat=false;
        dot=true;
        return;
    }
    if(source==jbuttons[2])
    {
        pressNumber("0");
        repeat=false;
        return;
    }
    if(source==jbuttons[3])
    {
        pressNumber("1");
        repeat=false;
        return;
    }
    if(source==jbuttons[4])
    {
        pressNumber("2");
        repeat=false;
        return;
    }
    if(source==jbuttons[6])
    {
        double temp=Math.cos(Double.parseDouble(dispresult.getText()));
        dispresult.setText(String.valueOf(temp));
        prev=true;
        repeat=false;
        dot=true;
        return;
    }
    if(source==jbuttons[7])
    {
        if(memoryArea.getText().equals("  M  "))
            dispresult.setText(String.valueOf(memery));
        prev=true;
        repeat=false;
        dot=true;
```

```
        return;
    }
    if(source==jbuttons[8])
    {
        pressNumber("3");
        repeat=false;
        return;
    }
    if(source==jbuttons[9])
    {
        pressNumber("4");
        repeat=false;
        return;
    }
    if(source==jbuttons[10])
    {
        pressNumber("5");
        repeat=false;
        return;
    }
    if(source==jbuttons[12])
    {
        double temp=Math.tan(Double.parseDouble(dispresult.getText()));
        dispresult.setText(String.valueOf(temp));
        prev=true;
        repeat=false;
        dot=true;
        return;
    }
    if(source==jbuttons[13])
    {
        memery=Double.parseDouble(dispresult.getText());
        if(memery !=0) memoryArea.setText("   M   ");
        prev=true;
        repeat=false;
        dot=true;
        return;
    }
    if(source==jbuttons[14])
    {
        pressNumber("6");
        repeat=false;
        return;
    }
```

```
if(source==jbuttons[15])
{
    pressNumber("7");
    repeat=false;
    return;
}
if(source==jbuttons[16])
{
    pressNumber("8");
    repeat=false;
    return;
}
if(source==jbuttons[18])
{
    double temp=-Double.parseDouble(dispresult.getText());
    dispresult.setText(String.valueOf(temp));
    prev=true;
    repeat=false;
    dot=true;
    return;
}
if(source==jbuttons[19])
{
    memery +=Double.parseDouble(dispresult.getText());
    if(memery !=0 ) memoryArea.setText("  M   ");
    prev=true;
    repeat=false;
    dot=true;
    return;
}
if(source==jbuttons[20])
{
    pressNumber("9");
    repeat=false;
    return;
}
if(source==jbuttons[21])
{
    if(dot){
        pressNumber(".");
        dot=false;
        repeat=false;
    }
    return;
```

```
    }
    if(source==jbuttons[22])
    {
        second=Double.parseDouble(dispresult.getText());
        dot=true;
        switch(secondsymbol)
        {
            case '*':
                second*=first;
                break;
            case '/':
                if(divide(second)) return;
                second=first/second;
        } //end of switch(secondsymbol)
        secondsymbol='\0';
        switch(firstsymbol)
        {
            case '+':
                result +=second;
                break;
            case '-':
                result -=second;
                break;
            case '*':
                result *=second;
                dispresult.setText(String.valueOf(result));
                break;
            case '/':
                if(divide(second)) return;
                result /=second;
        } //end of switch(firstsymbol)
        if(firstsymbol!='\0') dispresult.setText(String.valueOf(result));
        firstsymbol='\0';
        prev=true;
        repeat=false;
        return;
    }
    if(source==jbuttons[5])
    {
        dot=true;
        if(repeat)
        {
            firstsymbol='+';
            return;
```

```java
        }
    second=Double.parseDouble(dispresult.getText());
    switch(secondsymbol)
    {
        case '*':
            second*=first;
            break;
        case '/':
            if(divide(second)) return;
            second=first/second;
    } //end of switch(secondsymbol)
    secondsymbol='\0';
    switch(firstsymbol)
    {
        case '\0':
            result=second;
            firstsymbol='+';
            break;
        case '+':
            result +=second;
            dispresult.setText(String.valueOf(result));
            break;
        case '-':
            result -=second;
            firstsymbol='+';
            dispresult.setText(String.valueOf(result));
            break;
        case '*':
            result *=second;
            firstsymbol='+';
            dispresult.setText(String.valueOf(result));
            break;
        case '/':
            if(divide(second)) return;
            result /=second;
            firstsymbol='+';
            dispresult.setText(String.valueOf(result));
    } //end of switch(firstsymbol)
    prev=true;
    repeat=true;
    return;
}
if(source==jbuttons[11])
{
```

```java
            dot=true;
            if(repeat)
            {
                firstsymbol='-';
                return;
            }
            second=Double.parseDouble(dispresult.getText());
            switch(secondsymbol)
            {
                case '*':
                    second*=first;
                    break;
                case '/':
                    if(divide(second)) return;
                    second=first/second;
            }
            secondsymbol='\0';
            switch(firstsymbol)
            {
                case '\0':
                    result=second;
                    firstsymbol='-';
                    break;
                case '+':
                    result+=second;
                    firstsymbol='-';
                    dispresult.setText(String.valueOf(result));
                    break;
                case '-':
                    result-=second;
                    dispresult.setText(String.valueOf(result));
                    break;
                case '*':
                    result*=second;
                    firstsymbol='-';
                    dispresult.setText(String.valueOf(result));
                    break;
                case '/':
                    if(divide(second)) return;
                    result/=second;
                    firstsymbol='-';
                    dispresult.setText(String.valueOf(result));
            }
            prev=true;
```

```
        repeat=true;
        return;
    }
    if(source==jbuttons[17])
    {
        dot=true;
        if(repeat)
        {
            if(secondsymbol=='\0') firstsymbol='*';
            else secondsymbol='*';
            return;
        }
        second=Double.parseDouble(dispresult.getText());
        switch(secondsymbol)
        {
            case '\0':
                switch(firstsymbol)
                {
                    case '\0':
                        firstsymbol='*';
                        result=second;
                        break;
                    case '+':
                    case '-':
                        first=second;
                        secondsymbol='*';
                        break;
                    case '*':
                        result*=second;
                        dispresult.setText(String.valueOf(result));
                        break;
                    case '/':
                        if(divide(second)) return;
                        result/=second;
                        dispresult.setText(String.valueOf(result));
                        firstsymbol='*';
                }
                break;
            case '*':
                first*=second;
                dispresult.setText(String.valueOf(first));
                break;
            case '/':
                if(divide(second)) return;
```

```java
                    first /=second;
                    secondsymbol='*';
                    dispresult.setText(String.valueOf(first));
            }
            prev=true;
            repeat=true;
            return;
        }
        if(source==jbuttons[23])
        {
            dot=true;
            if(repeat)
            {
                if(secondsymbol=='\0') firstsymbol='/';
                else secondsymbol='/';
                return;
            }
            second=Double.parseDouble(dispresult.getText());
            switch(secondsymbol)
            {
                case '\0':
                    switch(firstsymbol)
                    {
                        case '\0':
                            firstsymbol='/';
                            result=second;
                            break;
                        case '+':
                        case '-':
                            first=second;
                            secondsymbol='/';
                            break;
                        case '*':
                            result*=second;
                            firstsymbol='/';
                            dispresult.setText(String.valueOf(result));
                            break;
                        case '/':
                            if(divide(second)) return;
                            result /=second;
                            dispresult.setText(String.valueOf(result));
                    } //end of switch(firstsymbol)
                    break;
                case '*':
```

```
                first *=second;
                secondsymbol='/';
                dispresult.setText(String.valueOf(first));
                break;
            case '/':
                if(divide(second)) return;
                first /=second;
                dispresult.setText(String.valueOf(first));
        } //end of switch(secondsymbol)
        prev=true;
        repeat=true;
        return;
    }
}
public static void main(String args[])
{
    MyComputer mc=new MyComputer();
}
}
```

19.4 程序说明

首先要设计计算器的界面。计算器窗口内的内容大多是按键,包括10个数字、3种常用的三角函数、5种常用运算符及等号和记忆区的标识。将这些按键全部定义为按钮。另外,还需要一个显示区,用来表示输入的值及最后计算的结果。显示区定义为文本区。

为了美观,将按键排列在矩形网格中,横成排竖成列。相应的语句如下:

```
GridLayout grid1;
grid1=new GridLayout(4, 6, 3, 3);           //放置按键
buttonField.setLayout(grid1);
```

使用循环,为每个按键添加事件监听器,并将按键添加到按键区域中。使用所选的字体显示按键上的标识。相应的语句如下:

```
for(int i=0;i<6;i++)
{
    for(int j=0;j<4;j++)
    {
        jbuttons[i*4+j]=new JButton(buttonname[i*4+j]);
        jbuttons[i*4+j].addActionListener(this);     //为各按钮添加事件监听器
        jbuttons[i*4+j].setFont(font);
        buttonField.add(jbuttons[i*4+j]);
    }
}
```

为按键添加了事件监听器后,需要实现其中的actionPerformed(ActionEvent e)方

法，以响应来自按钮的操作事件。主要的处理也是在这个方法内完成的。

首先定义事件源如下：

```
Object source=e.getSource();
```

之后根据事件源判定按下的是哪个键，然后进行相应的处理。例如，下面的语句处理正弦函数的计算。Java 提供了 Math 类，其中包含基本的数学操作，如计算指数、对数、平方根和三角函数等。

```
if(source==jbuttons[0])
{
    double temp=Math.sin(Double.parseDouble(dispresult.getText()));
    dispresult.setText(String.valueOf(temp));
    prev=true;
    repeat=false;
    dot=true;
    return;
}
```

计算的结果显示在显示区中，使用的方法是 setText()，方法的参数是要显示的内容。

第20章 实验7 用户界面设计

20.1 实验目的

本实验要求学生能够深入掌握界面设计方法及事件控制机制,能够实现绘图功能,具体的目的是:

(1) 了解 Java 提供的组件,根据用户界面的功能要求选择合适的组件。
(2) 了解布局管理器概念,能够按需求将组件组织成美观的界面。
(3) 能够控制各组件完成相应的功能。
(4) 能够实现绘图功能。
(5) 能够实现鼠标的各类响应。

20.2 实验内容

创建一个 Frame,窗口如图 20-1 所示。

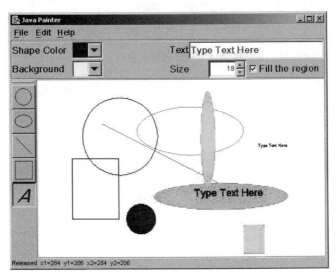

图 20-1 绘图窗口

20.3 参考答案

程序代码如下:

```
import javax.swing.*;
import java.awt.*;
import java.awt.event.*;
import javax.swing.event.*;
import java.math.*;
import java.io.*;
```

```java
import java.util.*;
import javax.swing.text.*;

class Draw extends JPanel
{
    MouseEvent first;                                    //上一个点
    MouseEvent last;                                     //椭圆、圆、矩形的开始点
    Graphics g;
    ObjectOutputStream out;
    ObjectInputStream in;
    ArrayList picture=new ArrayList();                   //存数据
    public int choose;                                   //选择的图形
    public Color fcolor;                                 //前景色
    public Color bcolor;                                 //背景色
    public int hollow;                                   //是否空心
    public int width;                                    //线的宽度
    public Draw()                                        //构造函数
    {
        addMouseListener(new Mouse1());
        //用于接收组件上的鼠标事件(按下、释放、单击、进入或离开)的侦听器接口
        addMouseMotionListener(new Mouse2());
        //用于接收组件上的鼠标移动事件的监听器接口
        choose=1;                                        //默认选择的是线
        hollow=0;                                        //默认是空心
        width=1;                                         //默认线宽是1
        fcolor=Color.black;                              //默认前景是黑色
        bcolor=Color.white;                              //默认背景是白色
        setbkcolor();
    }
    public void paint(Graphics graphics)                 //重绘函数
    {
        super.paint(graphics);
        setbkcolor();
        int i=0;
        while(i<picture.size())
        {
            MyShape shape=(MyShape)(picture.get(i++));
            shape.MyDraw(graphics);
        }
    }
    public void setbkcolor()                             //设置背景色
    {
        setBackground(bcolor);
    }
```

```java
    public void WriteToFile(ObjectOutputStream out)
    {
        try
        {
            out.writeObject(picture);
        }catch(Exception e)
        {
            JOptionPane.showMessageDialog(null,"SaveWrong",
                ":(",JOptionPane.INFORMATION_MESSAGE);
        }
    }
    public void ReadFromFile(ObjectInputStream in)
    {
        try
        {
            picture=(ArrayList)(in.readObject());
        }catch(Exception e)
        {
            JOptionPane.showMessageDialog(null,"OpenWrong",
                ":(",JOptionPane.INFORMATION_MESSAGE);
        }
        repaint();
    }
    public void addShape(MyShape shape)
    {
        picture.add(shape);
    }
    public void shapeDraw(MyShape shape)
    {
        Graphics gh=getGraphics();
        gh.setColor(fcolor);
        shape.MyDraw(gh);
    }
//******************鼠标响应******************
    class Mouse1 extends MouseAdapter
    {
        public void mousePressed(MouseEvent e)
        {
            first=e;
            last=e;
            if(choose==5)
            {
                MyText text=new MyText(e.getPoint(),DrawPen.textcontent,fcolor);
                addShape(text);
```

```
            shapeDraw(text);
        }
    }
    public void mouseReleased(MouseEvent e)
    {
        if(choose==1)
        {
            MyLine line=new MyLine(first.getPoint(),e.getPoint(),fcolor,
            width);
            addShape(line);
            shapeDraw(line);
        }
        if(choose==2)
        {
            int lx=(first.getPoint().x>e.getPoint().x)?
                    first.getPoint().x:e.getPoint().x;
            int ly=(first.getPoint().y>e.getPoint().y)?
                    first.getPoint().y:e.getPoint().y;
            int sx=(first.getPoint().x<e.getPoint().x)?
                    first.getPoint().x:e.getPoint().x;
            int sy=(first.getPoint().y<e.getPoint().y)?
                    first.getPoint().y:e.getPoint().y;
            for (int w=0; w<width; w++)
            {
                MyRect rect=new MyRect(sx+w,sy+w,lx-sx-2*w,ly-sy-2*w,
                fcolor,hollow);
                addShape(rect);
                shapeDraw(rect);
            }
        }
        if(choose==3)
        {
            int sx=(first.getPoint().x<e.getPoint().x)?
                    first.getPoint().x:e.getPoint().x;
            int sy=(first.getPoint().y<e.getPoint().y)?
                    first.getPoint().y:e.getPoint().y;
            MyCircle circle;
            for (int w=0; w<width; w++)
            {
                if(first.getPoint().x<e.getPoint().x &&
                   first.getPoint().y<e.getPoint().y)
                {
                    circle=new MyCircle(sx+w,sy+w,
                    Math.abs(e.getPoint().x-first.getPoint().x-2*w),
```

```
                Math.abs(e.getPoint().y-first.getPoint().y-2*w),
                    fcolor,hollow);
            }
            else if(first.getPoint().x<e.getPoint().x&&
                first.getPoint().y>=e.getPoint().y)
            {
                circle=new MyCircle(sx+w,sy+w,
                    Math.abs(e.getPoint().x-first.getPoint().x-2*w),
                    Math.abs(e.getPoint().y-first.getPoint().y+2*w),
                    fcolor,hollow);
            }
            else if(first.getPoint().x>=e.getPoint().x&&
                first.getPoint().y<e.getPoint().y)
            {
                circle=new MyCircle(sx+w,sy+w,
                    Math.abs(e.getPoint().x-first.getPoint().x+2*w),
                    Math.abs(e.getPoint().y-first.getPoint().y-2*w),
                    fcolor,hollow);
            }
            else
            {
                circle=new MyCircle(sx+w,sy+w,
                    Math.abs(e.getPoint().x-first.getPoint().x+2*w),
                    Math.abs(e.getPoint().y-first.getPoint().y+2*w),
                    fcolor,hollow);
            }
            addShape(circle);
            shapeDraw(circle);
        }
    }
    if(choose==4)
    {
        int x2=(e.getPoint().x-first.getPoint().x)*
            (e.getPoint().x-first.getPoint().x);
        int y2=(e.getPoint().y-first.getPoint().y)*
            (e.getPoint().y-first.getPoint().y);
        int r=(int)(Math.sqrt(x2+y2));
        for (int w=0; w<width; w++)
        {
            MyCircle circle=new MyCircle(first.getPoint().x-r+w,
                first.getPoint().y-r+w,2*r-2*w,2*r-2*w,fcolor,hollow);
            addShape(circle);
            shapeDraw(circle);
        }
```

```java
            }
        }
    }
    class Mouse2 extends MouseMotionAdapter
    {
        public void mouseDragged(MouseEvent e)
        {
            if(choose==1)
            {
                MyLine line=new MyLine(first.getPoint(),e.getPoint(),fcolor,
                width);
                addShape(line);
                shapeDraw(line);
                first=e;
            }
            if(choose==2)
            {
                repaint();
                int lx,ly,sx,sy;
                lx=(first.getPoint().x>e.getPoint().x)?first.getPoint().x:
                e.getPoint().x;
                ly=(first.getPoint().y>e.getPoint().y)?first.getPoint().y:
                e.getPoint().y;
                sx=(first.getPoint().x<e.getPoint().x)?first.getPoint().x:
                e.getPoint().x;
                sy=(first.getPoint().y<e.getPoint().y)?first.getPoint().y:
                e.getPoint().y;
                for (int w=0; w<width; w++)
                {
                    MyRect rect=new MyRect(sx+w,sy+w,lx-sx-2*w,ly-sy-2*w,
                    fcolor,hollow);
                    shapeDraw(rect);
                }
            }
            if(choose==3)
            {
                repaint();
                int sx,sy;
                sx=(first.getPoint().x<e.getPoint().x)?first.getPoint().x:
                e.getPoint().x;
                sy=(first.getPoint().y<e.getPoint().y)?first.getPoint().y:
                e.getPoint().y;
                MyCircle circle;
                for (int w=0; w<width; w++)
```

```java
                {
                    if(first.getPoint().x<e.getPoint().x &&
                        first.getPoint().y<e.getPoint().y)
                    {
                        circle=new MyCircle(sx+w,sy+w,
                            Math.abs(e.getPoint().x-first.getPoint().x-2*w),
                            Math.abs(e.getPoint().y-first.getPoint().y-2*w),
                            fcolor,hollow);
                    }
                    else if(first.getPoint().x<e.getPoint().x &&
                        first.getPoint().y>=e.getPoint().y)
                    {
                        circle=new MyCircle(sx+w,sy+w,
                            Math.abs(e.getPoint().x-first.getPoint().x-2*w),
                            Math.abs(e.getPoint().y-first.getPoint().y+2*w),
                            fcolor,hollow);
                    }
                    else if(first.getPoint().x>=e.getPoint().x &&
                        first.getPoint().y<e.getPoint().y)
                    {
                        circle=new MyCircle(sx+w,sy+w,
                            Math.abs(e.getPoint().x-first.getPoint().x+2*w),
                            Math.abs(e.getPoint().y-first.getPoint().y-2*w),
                            fcolor,hollow);
                    }
                    else
                    {
                        circle=new MyCircle(sx+w,sy+w,
                            Math.abs(e.getPoint().x-first.getPoint().x+2*w),
                            Math.abs(e.getPoint().y-first.getPoint().y+2*w),
                            fcolor,hollow);
                    }
                    shapeDraw(circle);
                }
        }
        if(choose==4)
        {
            repaint();
            int x2,y2,r;
            x2=(e.getPoint().x-first.getPoint().x)*
                (e.getPoint().x-first.getPoint().x);
            y2=(e.getPoint().y-first.getPoint().y)*
                (e.getPoint().y-first.getPoint().y);
            r=(int)(Math.sqrt(x2+y2));
```

```java
                for (int w=0; w<width; w++)
                {
                    MyCircle circle=new
                    MyCircle(first.getPoint().x-r+w,
                        first.getPoint().y-r+w,2 * r-2 * w,2 * r-2 * w,fcolor,
                        hollow);
                    shapeDraw(circle);
                }
            }
        }
    }
//******************图形类*******************
    abstract class MyShape implements Serializable
    {
        public abstract void MyDraw(Graphics graphics);
    }
    class MyLine extends MyShape implements Serializable
    {
        public MyLine(Point a,Point b,Color c,int w)
        {
            head=a;
            tail=b;
            color=c;
            width=w;
        }
        public void MyDraw(Graphics graphics)
        {
            Graphics2D g2d=(Graphics2D)graphics;
            graphics.setColor(color);
            g2d.setColor(color);
            if(width==1)
                graphics.drawLine(head.x,head.y,tail.x,tail.y);
            else
            {
                g2d.setStroke(new BasicStroke(width));
                g2d.drawLine(head.x,head.y,tail.x,tail.y);
            }
        }
        public Point head,tail;
        public Color color;
        public int width;
    }
    class MyRect extends MyShape implements Serializable
    {
```

```java
        public MyRect(int x1,int y1,int x2,int y2,Color c,int h)
        {
            ltx=x1;
            lty=y1;
            rbx=x2;
            rby=y2;
            color=c;
            hollow=h;
        }
        public void MyDraw(Graphics graphics)
        {
            graphics.setColor(color);
            if(hollow==0)
                graphics.drawRect(ltx,lty,rbx,rby);
            else
                graphics.fillRect(ltx,lty,rbx,rby);
        }
        public int ltx,lty,rbx,rby;
        public Color color;
        public int hollow;
}
class MyCircle extends MyShape implements Serializable
{
        public MyCircle(int x1,int y1,int x2,int y2,Color c,int h)
        {
            ltx=x1;
            lty=y1;
            rbx=x2;
            rby=y2;
            color=c;
            hollow=h;
        }
        public void MyDraw(Graphics graphics)
        {
            graphics.setColor(color);
            if(hollow==0)
                graphics.drawOval(ltx,lty,rbx,rby);
            else
                graphics.fillOval(ltx,lty,rbx,rby);
        }
        public int ltx,lty,rbx,rby;
        public Color color;
        public int hollow;
}
```

```java
class MyText extends MyShape implements Serializable
{
    public MyText(Point p,String t,Color c)
    {
        pos=p;
        text=t;
        color=c;
    }
    public void MyDraw(Graphics graphics)
    {
        graphics.setColor(color);
        graphics.drawString(text,pos.x,pos.y);
    }
    public Point pos;
    public String text;
    public Color color;
}
}
//******************主类********************
public class DrawPen extends JFrame
{
    public static String textcontent="Type Text Here";
    public Draw draw=new Draw();
    Controller control=new Controller();
    JMenu files=new JMenu("File");
    JMenu edits=new JMenu("Edit");
    JMenu helps=new JMenu("Help");
    JMenu fcolormenu=new JMenu("设置前景颜色");
    JMenu bcolormenu=new JMenu("设置背景颜色");
    JMenuItem[] filecmd={ new JMenuItem("打开文件"),
                                    new JMenuItem("保存文件"),
                                    new JMenuItem("退出")};
    JMenuItem[] fcolors={
        new JMenuItem("红色"),new JMenuItem("绿色"),
        new JMenuItem("蓝色"),new JMenuItem("黄色"),
        new JMenuItem("黑色"),new JMenuItem("颜色选择")};
    JMenuItem[] bcolors={
        new JMenuItem("红色"),new JMenuItem("绿色"),
        new JMenuItem("蓝色"),new JMenuItem("黄色"),
        new JMenuItem("白色"),new JMenuItem("颜色选择")
    };
//******************主函数********************
    public static void main(String[] args)
    {
```

```java
            try
            {
                UIManager.setLookAndFeel(UIManager.getSystemLookAndFeelClassName());
            }
            catch(Exception e)
            {
                e.printStackTrace(System.err);
            }
            DrawPen h=new DrawPen("画笔");
            h.setSize(800,600);
            h.setVisible(true);
        }
//*******************菜单下的一个JPanel*******************
    class Controller extends JPanel
    {
        JComboBox jcbfc=new JComboBox();                    //前景色 ComboBox
        JComboBox jcbbc=new JComboBox();                    //背景色 ComboBox
        JLabel jlfc=new JLabel();                           //前景色标签
        JLabel jlbc=new JLabel();                           //背景色标签
        JLabel jltext=new JLabel();                         //单行文本域标签
        JLabel jlsize=new JLabel();                         //线型标签
        JTextField jtf=new JTextField(textcontent);         //单行文本域
        Font font=new Font("Arial Rounded MT Bold",Font.PLAIN,15);
        JCheckBox jcb=new JCheckBox("Fill the region");    //圆、椭圆、矩形是否是实心的

        SpinnerNumberModel model=new SpinnerNumberModel(1,0,10,1);    //线型控制
        JSpinner jcbw=new JSpinner(model);
        String[] fcolors={"黑","红","绿","黄","蓝"};
        String[] bcolors={"白","红","绿","黄","蓝"};

        public Controller()
        {
            jlfc.setFont(font);
            jlbc.setFont(font);
            jltext.setFont(font);
            jlsize.setFont(font);
            jcb.setFont(font);
            jlfc.setText("Shape Color");
            jlbc.setText("Background");
            jltext.setText("Text");
            jlsize.setText("Size");
            JPanel upJP, downJP, tmpJP1, tmpJP2, tmpJP3, tmpJP4, tmpJP5;
            for(int i=0;i<5;i++)
            {
```

```
            jcbfc.addItem(fcolors[i]);
            jcbbc.addItem(bcolors[i]);
        }
        jcbfc.addActionListener(new ActionListener()
        {
            public void actionPerformed(ActionEvent e)
            {
                if(((JComboBox)e.getSource()).getSelectedItem()=="红")
                    draw.fcolor=Color.red;
                if(((JComboBox)e.getSource()).getSelectedItem()=="绿")
                    draw.fcolor=Color.green;
                if(((JComboBox)e.getSource()).getSelectedItem()=="黄")
                    draw.fcolor=Color.yellow;
                if(((JComboBox)e.getSource()).getSelectedItem()=="蓝")
                    draw.fcolor=Color.blue;
                if(((JComboBox)e.getSource()).getSelectedItem()=="黑")
                    draw.fcolor=Color.black;
            }
        });
        jcbbc.addActionListener(new ActionListener()
        {
            public void actionPerformed(ActionEvent e)
            {
                if(((JComboBox)e.getSource()).getSelectedItem()=="红")
                    draw.bcolor=Color.red;
                if(((JComboBox)e.getSource()).getSelectedItem()=="绿")
                    draw.bcolor=Color.green;
                if(((JComboBox)e.getSource()).getSelectedItem()=="黄")
                    draw.bcolor=Color.yellow;
                if(((JComboBox)e.getSource()).getSelectedItem()=="蓝")
                    draw.bcolor=Color.blue;
                if(((JComboBox)e.getSource()).getSelectedItem()=="白")
                    draw.bcolor=Color.white;
                draw.setbkcolor();
            }
        });
        jcbw.addChangeListener(new ChangeListener()
        {
            public void stateChanged(ChangeEvent e)
            {
                draw.width=Integer.parseInt(String.valueOf(jcbw.getValue()));
            }
        }
    );
```

```
jcb.addActionListener(new ActionListener()
{
    public void actionPerformed(ActionEvent e)
    {
        if(jcb.isSelected())
            draw.hollow=1;
        else
            draw.hollow=0;
    }
});
jtf.addActionListener(new ActionListener()
{
    public void actionPerformed(ActionEvent e)
    {
        textcontent=jtf.getText();
    }
});

upJP=new JPanel();                              //上一行
upJP.setLayout(new GridLayout(1,2,200,1));

tmpJP1=new JPanel();                            //放置前景颜色标签及组合框
tmpJP1.setLayout(new GridLayout(1,2,20,1));
tmpJP1.add(jlfc);                               //标签
tmpJP1.add(jcbfc);                              //组合框

tmpJP2=new JPanel();                            //放置文本标签及文本域
tmpJP2.setLayout(new GridLayout(1,2,20,1));

tmpJP2.add(jltext);
tmpJP2.add(jtf);
upJP.add(tmpJP1);
upJP.add(tmpJP2);

downJP=new JPanel();                            //下一行
downJP.setLayout(new GridLayout(1,2,200,1));

tmpJP3=new JPanel();                            //放置背景颜色标签及组合框
tmpJP3.setLayout(new GridLayout(1,2,20,1));
tmpJP3.add(jlbc);                               //标签
tmpJP3.add(jcbbc);                              //组合框

tmpJP4=new JPanel();                            //放置线型及组合框
tmpJP4.setLayout(new FlowLayout(FlowLayout.LEFT));
```

```
            tmpJP4.add(jlsize);
            tmpJP4.add(jcbw);

            tmpJP5=new JPanel();
            tmpJP5.setLayout(new GridLayout(1,2,10,1));
            tmpJP5.add(tmpJP4);
            tmpJP5.add(jcb);
            downJP.add(tmpJP3);
            downJP.add(tmpJP5);

            this.setLayout(new GridLayout(2,5));
            this.add(upJP);
            this.add(downJP);
        }
    }
    class setfcolor implements ActionListener
    {
        public void actionPerformed(ActionEvent e)
        {
            if(e.getActionCommand()=="红色")
                draw.fcolor=Color.red;
            if(e.getActionCommand()=="绿色")
                draw.fcolor=Color.green;
            if(e.getActionCommand()=="蓝色")
                draw.fcolor=Color.blue;
            if(e.getActionCommand()=="黄色")
                draw.fcolor=Color.yellow;
            if(e.getActionCommand()=="黑色")
                draw.fcolor=Color.black;
            if(e.getActionCommand()=="颜色选择")
            {
                Color c=null;
                c=JColorChooser.showDialog(DrawPen.this,"设置前景颜色",c);
                if(c!=null)
                    draw.fcolor=c;
            }
        }
    }
    class setbcolor implements ActionListener
    {
        public void actionPerformed(ActionEvent e)
        {
            if(e.getActionCommand()=="红色")
                draw.bcolor=Color.red;
```

```java
            if(e.getActionCommand()=="绿色")
                draw.bcolor=Color.green;
            if(e.getActionCommand()=="蓝色")
                draw.bcolor=Color.blue;
            if(e.getActionCommand()=="黄色")
                draw.bcolor=Color.yellow;
            if(e.getActionCommand()=="白色")
                draw.bcolor=Color.white;
            if(e.getActionCommand()=="颜色选择")
            {
                Color c=null;
                c=JColorChooser.showDialog(DrawPen.this,"设置背景颜色",c);
                if(c!=null)
                    draw.bcolor=c;
            }
            draw.setbkcolor();
        }
    }
    class setfile implements ActionListener
    {
        public void actionPerformed(ActionEvent e)
        {
            JFileChooser jfcdlg=new JFileChooser();
            if(e.getActionCommand()=="打开文件")
            {
                int i=jfcdlg.showOpenDialog(DrawPen.this);
                if(i==JFileChooser.APPROVE_OPTION)
                {
                    try
                    {
                        ObjectInputStream in=new ObjectInputStream(new
                            FileInputStream(jfcdlg.getSelectedFile()));
                        draw.ReadFromFile(in);
                        in.close();
                    }
                    catch(Exception ex)
                    {
                        JOptionPane.showMessageDialog(null,"打开失败",
                            "False",JOptionPane.INFORMATION_MESSAGE);
                    }
                }
            }
            if(e.getActionCommand()=="保存文件")
            {
```

```
                    int i=jfcdlg.showSaveDialog(DrawPen.this);
                    if(i==JFileChooser.APPROVE_OPTION)
                    {
                        try
                        {
                            ObjectOutputStream out=new ObjectOutputStream(new
                            FileOutputStream(jfcdlg.getSelectedFile()));
                            draw.WriteToFile(out);
                            out.close();
                        }
                        catch(Exception ex)
                        {
                            JOptionPane.showMessageDialog(null,"保存失败",
                                "False",JOptionPane.INFORMATION_MESSAGE);
                        }
                    }
                }
                if(e.getActionCommand()=="退出")
                {
                    System.exit(0);
                }
            }
        }
        setfile sf=new setfile();
        setfcolor sfc=new setfcolor();
        setbcolor sbc=new setbcolor();
//*******************构造函数*******************
        public DrawPen(String s)
        {
            super(s);
            for(int i=0;i<filecmd.length;i++)
            {
                filecmd[i].addActionListener(sf);
                files.add(filecmd[i]);
            }
            for(int i=0;i<fcolors.length;i++)
            {
                fcolors[i].addActionListener(sfc);
                fcolormenu.add(fcolors[i]);
            }
            for(int i=0;i<bcolors.length;i++)
            {
                bcolors[i].addActionListener(sbc);
                bcolormenu.add(bcolors[i]);
```

```java
}
JMenuBar mb=new JMenuBar();
files.setMnemonic(KeyEvent.VK_F);
edits.setMnemonic(KeyEvent.VK_E);
helps.setMnemonic(KeyEvent.VK_H);
mb.add(files);
mb.add(edits);
mb.add(helps);
mb.add(fcolormenu);
mb.add(bcolormenu);
setJMenuBar(mb);
JToolBar toolbar=new JToolBar();                    //工具条
toolBar.setOrientation(toolBar.VERTICAL);           //垂直放置
//工具条中各按钮对应的图案,保存在 images 目录下
JButton line=new JButton(new ImageIcon("images/line.gif"));
JButton rectangle=new JButton(new ImageIcon("images/rect.gif"));
JButton ellipse=new JButton(new ImageIcon("images/ellipse.gif"));
JButton circle=new JButton(new ImageIcon("images/circle.gif"));
JButton text=new JButton(new ImageIcon("images/text.gif"));
//给按钮注册事件监听器
line.addActionListener(new ActionListener()
{
    public void actionPerformed(ActionEvent e)
    {
        draw.choose=1;
    }
});
rectangle.addActionListener(new ActionListener()
{
    public void actionPerformed(ActionEvent e)
    {
        draw.choose=2;
    }
});
ellipse.addActionListener(new ActionListener()
{
    public void actionPerformed(ActionEvent e)
    {
        draw.choose=3;
    }
});
circle.addActionListener(new ActionListener()
{
    public void actionPerformed(ActionEvent e)
```

```
            {
                draw.choose=4;
            }
        });
        text.addActionListener(new ActionListener()
        {
            public void actionPerformed(ActionEvent e)
            {
                draw.choose=5;
            }
        });
        //将按钮加到工具条中
        toolBar.add(circle);
        toolBar.add(ellipse);
        toolBar.add(line);
        toolBar.add(rectangle);
        toolBar.add(text);
        Container cp=getContentPane();
        cp.add(draw);
        cp.add(BorderLayout.NORTH,control);
        cp.add(BorderLayout.WEST,toolBar);
    }
}
```

20.4　程序说明

首先要设计界面。界面中涉及的组件有：菜单、按钮、面板、组合框、单行文本区、复选框及数字选择(Spinner)组件等。

一般地，窗口中的菜单都配合工具条一起使用，使用工具条可以非常快速地定位到具体操作。常见的工具条位于菜单下方，本实验中要求工具条垂直放置在菜单下方画板的左侧。工具条中各按钮上有相应的图案，代表单击相应按钮后可画出的图形。通过 images 目录下的图形文件可以加载这些图案，images 目录包含在工作目录下，需要自己创建。

工具条的创建如下所示：

```
setJMenuBar(mb);
JToolBar toolBar=new JToolBar();                        //工具条
toolBar.setOrientation(toolBar.VERTICAL);               //垂直放置
JButton line=new JButton(new ImageIcon("images/line.gif"));
JButton rectangle=new JButton(new ImageIcon("images/rect.gif"));
JButton ellipse=new JButton(new ImageIcon("images/ellipse.gif"));
JButton circle=new JButton(new ImageIcon("images/circle.gif"));
JButton text=new JButton(new ImageIcon("images/text.gif"));
line.addActionListener(new ActionListener()
```

创建数字选择组件时，需要指定初始值、最小值、最大值及间隔。例如创建一个初值为 1、介于 0~10 之间增幅为 1 的数字选择组件的语句如下所示：

```
SpinnerNumberModel model=new SpinnerNumberModel(1, 0, 10, 1);
JSpinner jcbw=new JSpinner(model);
```

Java 提供了一个 Graphics 类，这是所有图形上下文的抽象基类。类中提供了多个画图方法，本实验中用到的有以下几个：

drawLine(int x1, int y1, int x2, int y2)，它的功能是在此图形上下文的坐标系统中，使用当前颜色在点(x1, y1)和(x2, y2)之间画一条线。

drawOval(int x, int y, int width, int height)，它的功能是绘制椭圆的边框。当 width 与 height 相等时，所画图形为圆形。

drawPolygon(int[] xPoints, int[] yPoints, int nPoints)，它的功能是绘制一个由横、纵坐标数组定义的闭合多边形。

drawRect(int x, int y, int width, int height)，它的功能是绘制指定矩形的边框。

fillOval(int x, int y, int width, int height)，它的功能是使用当前颜色填充外接指定矩形框内的椭圆。

fillPolygon(int[] xPoints, int[] yPoints, int nPoints)，它的功能是填充由 x 和 y 坐标数组定义的闭合多边形。

fillRect(int x, int y, int width, int height)，它的功能是填充指定的矩形。

程序中定义了抽象类 MyShape，类中有唯一的抽象方法 MyDraw(Graphics graphics)。该抽象类有 4 个子类，分别是 MyLine、MyRect、MyCircle 和 MyText。4 个子类中都需要实现抽象方法 MyDraw(Graphics graphics)，其中调用前面描述的几个预定义方法，实现具体的画图功能。

对鼠标事件进行控制时，需要给所控制的组件注册监听器。监听器都是接口，程序中需要实现接口中的全部抽象方法。为了方便程序编写，提高代码的可读性，Java 提供了抽象适配器类，程序中仅需要实现接口中感兴趣的方法，而忽略其他不需要的方法。

本实验中用到的适配器类有两个。一个是 MouseMotionAdapter，它接收鼠标移动或拖动事件。另一个是 MouseAdapter，它接收鼠标事件，包括鼠标何时按下、释放、单击、进入组件和离开组件等。

第21章 实验8 多线程练习

21.1 实验目的

本实验要求学生理解 Java 多线程的概念,包括线程的生命周期、优先级和调度等,能够编写多线程的程序,具体的目的是:

(1) 能够创建、管理、撤销线程。
(2) 能够设置线程的优先级。
(3) 能够控制线程的执行过程。

21.2 实验内容

编写一个多线程的控制程序,称为赛马程序。创建分别代表两匹马的两个线程,并将它们设置为高低不同的优先级,以进度条的形式显示赛马过程。

21.3 参考答案

程序的主界面在 MainFrame.java 文件中。程序代码如下:

```
package jthreadrace;

import java.awt.*;
import java.awt.event.*;
import javax.swing.*;

public class MainFrame extends JFrame
{
    private JPanel contentPane;
    private BorderLayout borderLayout1=new BorderLayout();
    private JLabel jLabel1=new JLabel();
    private JLabel jLabelleft=new JLabel("高优先级",SwingConstants.CENTER);
    private JLabel jLabelright=new JLabel("低优先级",SwingConstants.CENTER);
    private JPanel jLabelPanel=new JPanel();
    private JPanel jProgressPanel=new JPanel();
    private JProgressBar jProgressBarLeft=new JProgressBar();
                                    //进度条,用于显示线程进度
    private JProgressBar jProgressBarRight=new JProgressBar();
    private JButton jSwitchButton=new JButton();
    private JButton jExitButton=new JButton();
    private JPanel twoButton=new JPanel();

    private ProgressThread pThread1=null;
    //构造框架
```

```java
public MainFrame()
{
    enableEvents(AWTEvent.WINDOW_EVENT_MASK);
    try
    {
        jbInit();
    }catch(Exception e)
    {
        e.printStackTrace();
    }
}
//组件初始化
private void jbInit() throws Exception
{
    contentPane=(JPanel) this.getContentPane();
    jLabel1.setToolTipText("");
    jLabel1.setHorizontalAlignment(SwingConstants.CENTER);
    jLabelPanel.setLayout(new GridLayout(1,2));
    jLabelPanel.add(jLabelleft);
    jLabelPanel.add(jLabelright);
    contentPane.setLayout(borderLayout1);
    this.setSize(new Dimension(300, 200));
    this.setTitle("赛马线程");
    jProgressBarLeft.setOrientation(JProgressBar.VERTICAL);   //垂直显示进度条
    jProgressBarLeft.setMaximumSize(new Dimension(84, 32767));
                                                              //设置进度条的最大尺寸
    jProgressBarLeft.setPreferredSize(new Dimension(100, 148));   //首选大小
    jProgressBarLeft.setStringPainted(true);                  //呈现进度字符串
    jProgressBarRight.setOrientation(JProgressBar.VERTICAL);  //垂直显示进度条
    jProgressBarRight.setPreferredSize(new Dimension(100, 148));
    jProgressBarRight.setStringPainted(true);

    jProgressPanel.setLayout(new GridLayout(1,2,10,10));
    jProgressPanel.add(jProgressBarLeft);
    jProgressPanel.add(jProgressBarRight);

    jSwitchButton.setToolTipText("单击按钮开始赛马");
    jSwitchButton.setText("开始");
    jSwitchButton.addActionListener(new java.awt.event.ActionListener()
    {
        public void actionPerformed(ActionEvent e)            //注册事件监听器
        {
            jButton_actionPerformed(e);
        }
```

```
        });
        jExitButton.setText("退出");
        jExitButton.addActionListener(new java.awt.event.ActionListener()
        {
            public void actionPerformed(ActionEvent e)
            {
                jButton_actionPerformed(e);
            }
        });

        contentPane.add(jLabelPanel, BorderLayout.NORTH);
        contentPane.add(jProgressPanel, BorderLayout.CENTER);
        twoButton.add(jSwitchButton);
        twoButton.add(jExitButton);
        contentPane.add(twoButton, BorderLayout.SOUTH);
    }
    protected void processWindowEvent(WindowEvent e)
    {
        super.processWindowEvent(e);
        if(e.getID()==WindowEvent.WINDOW_CLOSING)
        {
            System.exit(0);
        }
    }

    void jButton_actionPerformed(ActionEvent e)                    //事件响应
    {
        if(((JButton)e.getSource()).getText().equals("开始"))
        {
            this.jSwitchButton.setText("停止");
            jSwitchButton.setToolTipText("单击按钮终止赛马");
            pThread1= new ProgressThread(this.jProgressBarLeft, Thread.MAX_
                PRIORITY);

            pThread1.start();
            ProgressThread pThread2=new ProgressThread( this.jProgressBarRight,
                    Thread.MIN_PRIORITY);
            pThread2.start();
        }
        else if (((JButton)e.getSource()).getText().equals("停止"))
        {
            this.jSwitchButton.setText("开始");
            jSwitchButton.setToolTipText("单击按钮开始赛马");
            this.pThread1.stopped=true;
```

```
            }
            else if(((JButton)e.getSource()).getText().equals("退出"))
            {
                System.exit(0);
            }
        }
    }
}
```

线程体代码在 ProgressThread.java 文件中。程序的代码如下：

```
package jthreadrace;
import javax.swing.*;

public class ProgressThread extends Thread
{
    JProgressBar pbar;
    static boolean stopped;

    public ProgressThread(JProgressBar pbar,int priority)
    {
        try
        {
            this.pbar=pbar;
            this.stopped=false;
            this.setPriority(priority);
        }catch(Exception err)
        {
            err.printStackTrace();
        }
    }

    public void run()
    {
        int min=0;
        int max=1000;
        this.pbar.setMinimum(min);
        this.pbar.setMaximum(max/10);
        this.pbar.setValue(min);

        for(int i=min; i<=max; i++)
        {
            if(stopped) break;
            else
            {
                this.pbar.setValue((int)(i/10));
```

```
            this.pbar.setString(String.valueOf(i));
            try
            {
                Thread.sleep(10);
            }catch(Exception err)
            {
                err.printStackTrace();
            }
        }
      }
   }
}
```

测试程序在 JThreadRace.java 文件中。程序代码如下：

```
package jthreadrace;

import javax.swing.UIManager;
import java.awt.*;

public class JThreadRace
{
    private boolean packFrame=false;

    //创建应用程序
    public JThreadRace()
    {
        MainFrame frame=new MainFrame();
        if (packFrame)
        {
            frame.pack();
        }
        else
        {
            frame.validate();
        }
        Dimension screenSize=Toolkit.getDefaultToolkit().getScreenSize();
        Dimension frameSize=frame.getSize();
        if (frameSize.height>screenSize.height)
        {
            frameSize.height=screenSize.height;
        }
        if (frameSize.width>screenSize.width)
        {
            frameSize.width=screenSize.width;
```

```
        }
        frame.setLocation((screenSize.width-frameSize.width)/2,
            (screenSize.height-frameSize.height)/2);
        frame.setVisible(true);
    }
    public static void main(String[] args)
    {
        new JThreadRace();
    }
}
```

21.4 程序说明

线程的主要控制都在线程体中,编写 run()函数完成线程体。

第 22 章 实验 9 文件读写练习

22.1 实验目的

本实验要求学生理解 Java 数据流概念,理解 Java 流类的层次结构,能够访问文件,具体的目的是:

(1) 能够创建、读写和更新文件。
(2) 能够使用 ObjectInputStream 和 ObjectOutputStream 流。
(3) 熟悉顺序存取文件和随机存取文件的处理。

22.2 实验内容

编写一个个人通信录程序,具有如下的功能:

- 定义一个类,包含姓名、邮政编码、通信地址等成员变量。
- 查找:根据姓名在文件中查找个人信息,如果找到则显示出来。
- 添加:向文件中写个人信息。
- 通信录界面如图 22-1 所示。

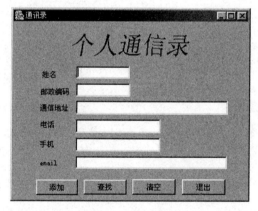

图 22-1 个人通信录

22.3 参考答案

程序代码如下:

```
import javax.swing.*;
import java.awt.*;
import java.awt.event.*;
import java.io.*;
import java.util.*;

public class Communication extends JFrame
```

```java
{
    //定义界面中必要的组件,包括标签、文本域、按钮等
    JLabel title=new JLabel("个人通信录");
    JLabel name=new JLabel("姓名");
    JLabel zip=new JLabel("邮政编码");
    JLabel address=new JLabel("通信地址");
    JLabel telephone=new JLabel("电话");
    JLabel mobile=new JLabel("手机");
    JLabel email=new JLabel("email");
    JTextField nametext=new JTextField();
    JTextField ziptext=new JTextField();
    JTextField addtext=new JTextField();
    JTextField teltext=new JTextField();
    JTextField mobtext=new JTextField();
    JTextField emailtext=new JTextField();
    Font font=new Font("TimersRoman", Font.ITALIC,40);      //使用的字体
    JButton add=new JButton("添加");
    JButton find=new JButton("查找");
    JButton clear=new JButton("清空");
    JButton exit=new JButton("退出");
    ArrayList NameCardArray=new ArrayList();
    Communication(String s)
    {
        super(s);
        Container cp=getContentPane();
        cp.setLayout(null);

        //给各文本输入域设置边框
        nametext.setBorder(BorderFactory.createLoweredBevelBorder());
        ziptext.setBorder(BorderFactory.createLoweredBevelBorder());
        addtext.setBorder(BorderFactory.createLoweredBevelBorder());
        teltext.setBorder(BorderFactory.createLoweredBevelBorder());
        mobtext.setBorder(BorderFactory.createLoweredBevelBorder());
        emailtext.setBorder(BorderFactory.createLoweredBevelBorder());

        //给各按钮设置边框
        add.setBorder(BorderFactory.createLoweredBevelBorder());
        find.setBorder(BorderFactory.createLoweredBevelBorder());
        clear.setBorder(BorderFactory.createLoweredBevelBorder());
        exit.setBorder(BorderFactory.createLoweredBevelBorder());

        title.setFont(font);                    //设置文本域使用的字体
        //设置各组件的大小
        title.setBounds(130, 20, 300, 60);
```

```java
            name.setBounds(50, 100, 75, 20);
            nametext.setBounds(150, 100, 100, 20);
            zip.setBounds(50, 140, 75, 20);
            ziptext.setBounds(150, 140, 100, 20);
            address.setBounds(50, 180, 75, 20);
            addtext.setBounds(150, 180, 250, 20);
            telephone.setBounds(50, 220, 75, 20);
            teltext.setBounds(150, 220, 150, 20);
            mobile.setBounds(50, 260, 75, 20);
            mobtext.setBounds(150, 260, 150, 20);
            email.setBounds(50, 300, 75, 20);
            emailtext.setBounds(150, 300, 250, 20);
            //设置按钮的位置
            add.setBounds(50, 360, 75, 25);
            find.setBounds(150, 360, 75, 25);
            clear.setBounds(250, 360, 75, 25);
            exit.setBounds(350, 360, 75, 25);
            add.addActionListener(new ActionListener()        //给添加按钮注册监听器
            {
                public void actionPerformed(ActionEvent e)
                {
                    if(nametext.getText().equalsIgnoreCase(""))
                    {
                        JOptionPane.showMessageDialog(null,"无法添加名字为空的记录",
                            "Message", JOptionPane.INFORMATION_MESSAGE);
                        nametext.setText("");
                        ziptext.setText("");
                        addtext.setText("");
                        teltext.setText("");
                        mobtext.setText("");
                        emailtext.setText("");
                        return;
                    }
                    //从各文本域中获取新数据中
                    Note note=new Note();
                    note.name=nametext.getText();
                    note.zip=ziptext.getText();
                    note.address=addtext.getText();
                    note.telephone=teltext.getText();
                    note.mobile=mobtext.getText();
                    note.email=emailtext.getText();
                    try {
                        ObjectInputStream in=new ObjectInputStream(new
                            FileInputStream("note.dat"));
```

```java
            //从原文件中读入已有的数据
            NameCardArray=(ArrayList)in.readObject();
            in.close();
        } catch(Exception ex) {}
        try {
            ObjectOutputStream out=new ObjectOutputStream(new
                FileOutputStream("note.dat", true));
            Note temp=new Note();
            int i;
            //判定是否存在相同的记录
            for(i=0; i<NameCardArray.size(); i++)
            {
                temp=(Note)NameCardArray.get(i);
                if(temp.name.equalsIgnoreCase(nametext.getText()))
                    break;
            }
            if(NameCardArray.size() !=0 && i !=NameCardArray.size())
            {
                JOptionPane.showMessageDialog(null,"已经存在此记录",
                    "Message",JOptionPane.INFORMATION_MESSAGE);
            }
            else
            {
                NameCardArray.add(note);
                //回写数据
                out.writeObject(NameCardArray);
            }
            out.close();
        } catch(Exception ex) {}
        nametext.setText("");
        ziptext.setText("");
        addtext.setText("");
        teltext.setText("");
        mobtext.setText("");
        emailtext.setText("");
    }
});
find.addActionListener(new ActionListener()          //给查找按钮注册监听器
{
    public void actionPerformed(ActionEvent e)
    {
        try {
            ObjectInputStream in=new ObjectInputStream(new
                FileInputStream("note.dat"));
```

```java
            NameCardArray=(ArrayList)in.readObject();
            in.close();
        } catch(Exception ex) {}
        Note temp=new Note();
        int i;
        for(i=0; i<NameCardArray.size(); i++)
        {
            temp=(Note) NameCardArray.get(i);
            if(temp.name.equalsIgnoreCase(nametext.getText()))
                break;
        }
        if(NameCardArray.size() !=0 && i !=NameCardArray.size())
        {
            ziptext.setText(temp.zip);
            addtext.setText(temp.address);
            teltext.setText(temp.telephone);
            mobtext.setText(temp.mobile);
            emailtext.setText(temp.email);
        }
        else
        {
            nametext.setText("");
            ziptext.setText("");
            addtext.setText("");
            teltext.setText("");
            mobtext.setText("");
            emailtext.setText("");
            JOptionPane.showMessageDialog(null, "无此记录", "Message",
                JOptionPane.INFORMATION_MESSAGE);
        }
    }
});
clear.addActionListener(new ActionListener()      //给清空按钮注册监听器
{
    public void actionPerformed(ActionEvent e)
    {
        try
        {
            ObjectOutputStream out=new ObjectOutputStream(new
                FileOutputStream("note.dat"));
            NameCardArray.clear();
            out.close();
        } catch(Exception ex) {}
        nametext.setText("");
```

```
                ziptext.setText("");
                addtext.setText("");
                teltext.setText("");
                mobtext.setText("");
                emailtext.setText("");
            }
        });
        exit.addActionListener(new ActionListener()        //给退出按钮注册监听器
        {
            public void actionPerformed(ActionEvent e)
            {
                System.exit(1);
            }
        });
        //布置组件
        cp.add(title);
        cp.add(name);
        cp.add(zip);
        cp.add(address);
        cp.add(telephone);
        cp.add(mobile);
        cp.add(email);
        cp.add(nametext);
        cp.add(ziptext);
        cp.add(addtext);
        cp.add(teltext);
        cp.add(mobtext);
        cp.add(emailtext);
        cp.add(add);
        cp.add(find);
        cp.add(clear);
        cp.add(exit);
    }
    public static void main(String[] args)
    {
        Communication com=new Communication("通信录");
        com.addWindowListener(new WindowAdapter()
        {
            public void windowClosing( WindowEvent e )
            {
                System.exit(0);
            }
        });
        com.setSize(480,460);
```

```
            com.setVisible(true);
        }
    }

    class Note implements Serializable
    {
        public String name;
        public String zip;
        public String address;
        public String telephone;
        public String mobile;
        public String email;
        public Note() {}
    }
```

22.4 程序说明

程序中定义了 ArrayList 类的数组 NameCardArray。ArrayList 类实现的是大小可变的数组,从文件中读入的数据先保存在数组 NameCardArray 中,添加或是查找都要在该数组中进行。最后,再将数组中的内容写回文件中保存。